The Beginner's Guide to Engineering:
Electrical Engineering

quantum scientific publishing

The Beginner's Guide to Engineering:
Electrical Engineering

Mary Ellen Latschar

quantum scientific publishing

The Beginner's Guide to Engineering: Electrical Engineering

ISBN-13: 978-1492986652
ISBN-10: 1492986658

Published by quantum scientific publishing

Pittsburgh, PA | Copyright © 2013

All rights reserved. Permission in writing must be obtained from the publisher before any part of this work may be reproduced or transmitted in any form, including photocopying and recording.

Cover design by Scott Sheariss
Cover photo courtesy of Carl Silver

Unit One

Section 1.1 – Life without Electricity   8

Section 1.2 – Definition of Electricity   11

Section 1.3 – Electrical Units of Measurement   17

Section 1.4 – The Generation of Electricity   23

Section 1.5 – Power, Energy and Work   27

Section 1.6 – What is an Electrical Circuit?   31

Section 1.7 – Conductors, Insulators, Resistors, Variable Resistors and Capacitors   37

Section 1.8 – Batteries and Their Uses   41

Section 1.9 – The Difference between Alternating and Direct Current   45

Section 1.10 – Definition of Static Electricity   49

Section 1.11 – The Light Bulb   53

Section 1.12 – Magnets and Magnetic Fields   59

Section 1.13 – The Generation of Heat by Electricity   61

Section 1.14 – Grounding   63

Section 1.15 – Circuit Breakers   65

Unit Two

Section 2.1 – Electronic Symbols   70

Section 2.2 – Ohm's Law   75

Section 2.3 – Basic Circuit Concepts – Series Circuits   79

Section 2.4 – Basic Circuit Concepts – Parallel Circuits   83

Section 2.5 – Basic Circuit Concepts – Combination Circuits   85

Section 2.6 – Basic Circuit Concepts – Kirchhoff's Current Law   87

Section 2.7 – Basic Circuit Concepts – Kirchhoff's Voltage Law   91

Section 2.8 – Basic Circuit Concepts – Solving Circuits   95

Section 2.9 – Basic Circuit Concepts – Thévenin's Theorem   101

Section 2.10 – Basic Circuits Concepts–Norton's Theorem   105

Section 2.11 – Basic Circuit Concepts – Superposition Theorem   107

Section 2.12 – Basic Circuit Concepts – Millman's Theorem   115

Section 2.13 – Measuring Instruments   119

Section 2.14 – The Wheatstone Bridge   123

Section 2.15 – An In-Depth Review of Circuit Concepts   125

Unit Three

Section 3.1 – How Things Work – On/Off Switches   129

Section 3.2 – How Things Work – Lights and Lighting   137

Section 3.3 – How Things Work – Portable Lights   149

Section 3.4 – How Things Work – Car Batteries   157

Section 3.5 – How Things Work – Water Pump in an Internal Combustion Engine of an Automobile   167

Section 3.6 – How Things Work – Ceiling Fans   173

Section 3.7 – How Things Work – Portable Electric Heaters   179

Section 3.8 – How Things Work – Microwave Ovens   189

Section 3.9 – How Things Work – Wireless Remote Controllers   195

Section 3.10 – How Things Work – Safety Sensors (Carbon Monoxide Sensors)   203

Section 3.11 – How Things Work – Direct Current Motors   211

Section 3.12 – How Things Work – Alternating Current Motors   225

Section 3.13 – U.S. and International Electrical Standards   233

Section 3.14 – How Things Work -Power Transmission and Alternative Sources   243

Section 3.15 – Applications of Electrical Engineering   253

Appendix

Unit One Answer Key   268

Unit Two Answer Key   273

Unit Three Answer Key   278

Unit One

Section 1.1 – Life without Electricity 8

Section 1.2 – Definition of Electricity 11

Section 1.3 – Electrical Units of Measurement 17

Section 1.4 – The Generation of Electricity 23

Section 1.5 – Power, Energy and Work 27

Section 1.6 – What is an Electrical Circuit? 31

Section 1.7 – Conductors, Insulators, Resistors, Variable Resistors and Capacitors 37

Section 1.8 – Batteries and Their Uses 41

Section 1.9 – The Difference between Alternating and Direct Current 45

Section 1.10 – Definition of Static Electricity 49

Section 1.11 – The Light Bulb 53

Section 1.12 – Magnets and Magnetic Fields 59

Section 1.13 – The Generation of Heat by Electricity 61

Section 1.14 – Grounding 63

Section 1.15 – Circuit Breakers 65

Section 1.1 – Life Without Electricity

Section Objective

- Describe why electricity is important

Electricity and You

What would your life be without electricity? You would wake up in the morning without an alarm clock, wash yourself with cold water in a bucket, get dressed and go into the kitchen for breakfast. There would be no microwave. There would be no toaster. There would be no stove unless you had a wood-burning stove. There would probably be an open hearth for a fire. You would either eat food that was raw and uncooked or cooked over a fire.

> Thomas Edison (1847-1931)
>
> Holds 1,093 patents, a world record for his time. He invented the first practical electric light bulb, phonograph, and early movie projectors called kinetoscopes,

Woman in Puritan attire seated at a spinning wheel by a fireplace (1906.)
Image courtesy of the Library of Congress

You would not be able to call your best friend because there would be no phones. Cell phones would not exist. There would be no bus or car to take you to school. You would have to walk to school unless you had a horse. Now having a horse might sound like a lot of fun, but think about it. You have to feed a horse. You have to wash him and groom him. You also have to muck out his stall. In the wind, the snow, and the rain, both you and the horse would be exposed to the elements.

Once you are in school, there would be no central air-conditioning during the warm months and no central heating during the cold. There would be no electric typewriters, no computers, no Internet and NO CALCULATORS! You would have to write out all of your lessons in long hand and do all of your calculations manually. However, you might have use of an abacus to help you calculate numbers!

An abacus from a Danish school. Early 20th Century.

There would be no electric lights. All lighting would be natural lighting from the sun or from something like candles. There would be no flashlights because batteries would not exist.

When you went home after school, you would have to do your share of chores. Everyone would have to work together in order to survive. When you are done, there would be no TV to watch and no video games to play. Wii™ would not exist.

Society as we know it would not be the same. Many conveniences that we take for granted would not exist without electricity. Numerous jobs would not exist without electricity. There would be no power plants, no automotive plants, no computer companies, no video companies, and no telephone companies to name a few. Who do you know that would not have a job if there were no electricity?

A Little History

You don't have to go back too many generations to find one of your ancestors who lived without electricity. Electricity existed, but we had to learn how to harness and use it. The ancient Greeks found that when they rubbed amber it attracted straw and when they rubbed lodestone it attracted iron. While they found it fascinating, they didn't understand why this attraction occurred. The Chinese were able to magnetize steel from a lodestone possibly leading to the discovery of the compass as early as the third century A.D.

Many experiments were performed throughout the ages by various scholars in civilized countries. Otto von Guericke (1660) was credited with building the first electrical machine which studied the phenomenon of the attraction by static charges. In 1745 Pieter van Musschenbroek developed the Leyden jar that "stored" static electricity. Thomas Edison was able to demonstrate a working electric light in 1879. Three years later in 1882, the electric light was commercially available after Edison established a working central electricity generating station on Pearl Street in New York City.

> Otto von Guericke (1602-1686) Proved that a vacuum could exist. He invented a machine that produced static electricity in 1660.

> Pieter van Musschenbroek Invented the first device to store static electricity, called a Leyden Jar. The Leyden Jar was the original capacitor, which is a device that stores and releases electrical charge.

A Temporary Loss of Electricity

It has become common upon occasion to experience a temporary loss of electricity due to inclement weather or an equipment failure. A heavy wind storm blows a heavy tree limb down on a power line interrupting the flow of electricity and causes a power outage. Upon rare occasion, the utility company experiences a problem and power is cut off to your neighborhood. When this happens, the loss of power is usually only for a short period of time. But in a major disaster, that time could become days or even weeks. Remember the problems in New Orleans after Hurricane Katrina? If you weren't there, it is hard to imagine what it was like.

If you are having a short term power failure, it is usually just annoying. You get out flashlights and battery operated lanterns. Then you complain about the fact that you can't watch TV or play a video game. What a bummer!

What doesn't work in a power outage? With the use of batteries and generators, some things still function. Flashlights and lanterns provide light. Battery operated radios provide music and communication with the outside world. Microwaves will not work without power, but an outdoor gas grill should. Wood burning fireplaces would work as would any wood burning stove. Your ipod would work as long as the battery lasted. Generators kick in to power critical emergency needs (hospitals, emergency lighting, etc.)

Doesn't sound too bad if it is only lasts a few hours, does it? But what if the power outage lasted for several days? The situation starts to be uncomfortable. Air conditioning, furnaces, cooking appliances, computers, and the internet would not function. What if we run out of batteries?

Power has become an integral part of our existence. Lives would be lost because hospitals could not function. Travel would be very difficult and time consuming. Nearly all of our activities would be impacted.

Concept Reinforcement

1. Describe what a Monday would be like for you if you woke up to a total power outage, but school was not cancelled. Go through your day in detail and identify anything that requires electricity.

2. Think about your home and school. What would you need if the power were to go out for more than 3 days?

3. How could you prepare for a power outage at home? At school? What items should you have just in case this outage occurs?

4. Identify 3 different modes of transportation that would work without power.

Section 1.2 – Definition of Electricity

Section Objective

- Explain electricity and its components

Electricity and Its Components

To understand electricity, we have to understand the basic construction of an atom. An atom is considered the basic building block of the universe because all matter is made up of atoms. Elements are materials that are made of a single atom. For example, elemental copper is made of copper atoms. Elemental gold is made of gold atoms. Take a look at part of the periodic table. The first 30 elements are listed below. Each atom (element) has a unique atomic number, a name, and a symbol used to identify the element. If you look at the periodic table you, will see the number of valence electrons associated with each element or atom. The valence electrons are important to understanding the concept of electricity, which will be explained in this section.

	First 30 Elements of the Periodic Table		
Atomic Number	Name	Valence Electrons	Symbol
1	Hydrogen	1	H
2	Helium	2	He
3	Lithium	1	Li
4	Beryllium	2	Be
5	Boron	3	B
6	Carbon	4	C
7	Nitrogen	5	N
8	Oxygen	6	O
9	Fluorine	7	F
10	Neon	8	Ne
11	Sodium	1	Na
12	Magnesium	2	Ma
13	Aluminum	3	Al
14	Silicon	4	Si
15	Phosphorus	5	P
16	Sulfur	6	S
17	Chlorine	7	Cl
18	Argon	8	A
19	Potassium	1	K
20	Calcium	2	Ca
21	Scandium	2	Sc
22	Titanium	2	Ti
23	Vanadium	2	V
24	Chromium	1	Cr
25	Manganese	2	Mn
26	Iron	2	Fe
27	Cobalt	2	Co
28	Nickel	2	Ni
29	Copper	1	Cu
30	Zinc	2	Zn

The Atom

The three principal parts of an atom are the electron, neutron and proton. Neutrons and protons combine to form the nucleus or center of the atom. Electrons orbit the nucleus of the atom. This orbit is often represented as a defined orbit. The orbital path is called a shell. However, scientists are only able to estimate the location of an electron relative to the nucleus of the atom. Element number 1 is hydrogen. Hydrogen has one proton and one electron. The proton forms the nucleus. The proton has a positive charge and the electron has a negative charge. The image below shows the two principal parts of the hydrogen atom.

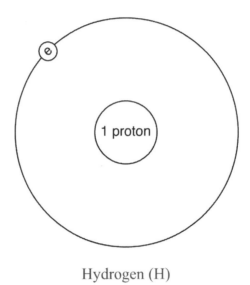

Hydrogen (H)

The atomic number is equal to the number of protons for each element. The nucleus may or may not contain the same number of protons and neutrons. In a balanced atom, the number of protons equals the number of atoms.

The nucleus of lithium whose atomic number is three contains three protons and four neutrons. Lithium also has 3 electrons. Notice that the 3rd electron moves around the nucleus in a larger orbit than the first 2 electrons.

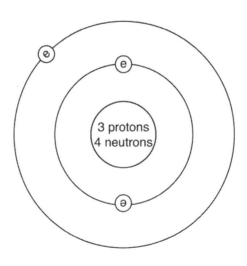

Lithium (Li)

Copper which is number 29 in the periodic table contains 29 protons and 35 neutrons. Copper also contains 29 electrons.

The electrons orbit around the nucleus. The first orbit and the one closest to the nucleus can hold no more than 2 electrons.

The second orbit can hold no more than 8 electrons

The third orbit can hold no more than 18 electrons.

The fourth orbit can hold no more than 32 electrons

As you can see in picture of the copper atom, copper has 2 electrons in the first orbit, 8 electrons in the second orbit, 18 electrons in the third orbit and only one in the fourth (outer orbit). This makes a total of 29 electrons.

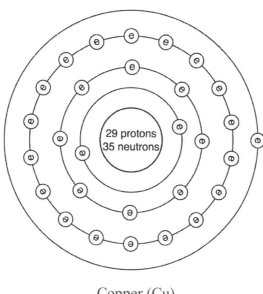

Copper (Cu)

Valence Electrons

We are interested in the electrons in the outermost orbit (shell). The electrons in the outer orbit are called valence electrons. Atoms that do not have completely filled outer orbits are unstable and want to move toward stability. An atom with a completely filled atomic valence is a stable atom.

The electrons in the outermost orbit are not as stable as those in the inner orbits. The electrons in the outermost orbit are furthest away from the nucleus. Thus when they are disturbed or energized, they are more apt to leave their orbit (shell) and move on.

Atoms with fewer valence electrons tend to move toward a stable state. This is one of the key aspects of electricity. Electricity is generated as electrons are transferred from one atom to another. Copper has only one valence electron in its outer orbit, meaning that it will easily give up a valence electron, which will result in a release of energy. To understand what this has to do with electricity, you need to understand the following two laws of physics.

The Law of Charges

The Law of Charges states that opposite charges attract and like charges repel. If you put two magnets together, one end attracts and the other repels. Electrons are negative and attracted to positive protons.

The Law of Centrifugal Force

The Law of Centrifugal Force states that a rotating or spinning object will pull away from its center the faster it spins. The force the object exerts on the center will increase as the rotational speed increases. If you tie a ball to a string and spin it around you, the faster you turn, the straighter the string will become. You will feel the force of the ball pulling away from you. If you slow down, the force on the rotating object will lessen and the ball will start to drop. Electrons orbit around the nucleus of an atom and the valence electrons are in the outermost orbit around the nucleus. Centrifugal force is an outward force associated with rotation.

The Law of Charges is binding the electrons to the nucleus while Centrifugal Force is pulling them away.

Electrical Current is the Flow of Electrons

Electrical current is the result of the movement of valence electrons between atoms. This movement releases energy. When an atom collides with another atom, a valence electron orbiting in the outermost orbit of the first atom bumps into a valence electron orbiting around the nucleus of the second atom. When this happens, the valence electron in the outer orbit of the second atom is knocked out of its orbit, releasing energy. The valence electron is released from the first atom and replaces the valence electron it hit, while the valence electron that was bumped out of its orbit moves on to strike another valence electron in another atom. This movement of electrons creates electrical flow.

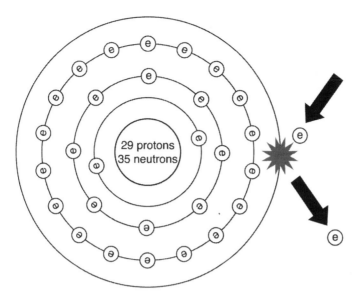

Electron of one atom knocks electron of another out of orbit

If an electron hits the valence electron, the first electron will push the second electron out of its current orbit and transfer the bulk of its energy to that electron. Because it will have lost its excess energy, the first electron will take the second electron's place in orbit. The second electron will then go on to hit a third electron in another atom.

This process continues and the result is the movement of electrons. The movement of electrons is electrical current.

However, not all of the striking electron's energy is transferred to the electron it hit. Some of the energy is lost as heat. Sometimes when you touch a wire that has electricity (current) flowing through it, you are feeling the heat that is lost when electrons collide and transfer energy. Beware of wires that become too warm to touch. Wires that become VERY hot could potentially cause a fire.

Concept Reinforcement

1. Explain the Law of Change and the Law of Centrifugal Force.

2. List the components of an atom.

3. Explain a valence electron.

4. Explain electrical current.

Section 1.3 – Electrical Units of Measurement

Section Objective

- Identify units of measurements used for electricity

Coulombs – the Unit for Electrical Charge

Charles Augustin de Coulomb, a French scientist in the 1700s, experimented with electrostatic charges and developed an equation describing the electrostatic force between electric charges. Coulomb's law is based on the attraction and repulsion between bodies of the same and opposite electrical charge and the distance between them.

Charles-Augustin de Coulomb

Coulomb's Law of Electrostatic Charges states that the force of electrostatic attraction or repulsion is directly proportional to the product of the two charges and inversely proportional to the square of the distance between them. Basically, it is a means by which to measure electrical charge. Coulomb is a unit representing electric charge. One coulomb equals the amount of electric charge on 6.25×10^{18} electrons.

> **Coulomb's Law**
> The force of electrostatic attraction or repulsion is directly proportional to the product of the two charges and inversely proportional to the square of the distance between them.
>
> The unit of the Coulomb is written as C.
>
> $1\ C = 6.25 \times 10^{18}$ electrons

Amps – the Unit of Electric Current

André-Marie Ampère (1775-1836), a French physicist, founded the science of electrodynamics now known as electromagnetism. He built an instrument, the galvanometer, to measure the flow of electricity.

André-Marie Ampère

> **Ampere**
>
> An ampere is the amount of current that flows through a point in a wire in one second.
>
> The ampere is represented by amp or the letter A.

The ampere (amp) is a measurement of the amount of electricity that flows through a circuit. An amp, or ampere, is represented by "A." The symbol "I" is frequently used to represent the flow of electrical current in equations. Refer to Ohm's Law described in this chapter. One ampere of current occurs when one coulomb of electrical charge passes a point in a wire in one second.

Volt – the Unit of "Energy"

Voltage is the force that pushes electrons through a wire. Voltage is also defined as electromotive force (EMF). A volt causes one coulomb to produce one joule of work. Voltage is the difference in electrical potential between two points in a circuit and is measured in volts (represented by "V"). Voltage does not flow through a wire, it pushes current (electrons) through a wire.

The volt is named for Alessandro Volta, an Italian physicist who performed pioneering work in electricity. He is best known for developing the voltaic pile, which was later developed into the modern battery.

Alessandro Volta

Watt – the Unit of Power

Wattage is the amount of power that is used in an electric circuit. The watt was named to honor the scientist James Watt, who developed the steam engine. The steam engine revolutionized society, speeding the transition from an agricultural society to an industrial society.

James Watt

Wattage is proportional to the amount of current flowing through an electric circuit and the amount of voltage in that circuit. Wattage is a measure of power. Power is the amount of current times the voltage level at a given point. The unit of measurement is the watt (represented by "P" or "W"). Electric power is the rate electrical energy is transferred by an electric circuit. In order for power (watts) to exist, electrical energy must be converted into heat or mechanical energy. The relationship between power, energy and work will be discussed in a later chapter.

Ohm – the Unit of Resistance (the Load)

An ohm is the amount of resistance that allows 1 amp of current to flow when 1 volt is applied to the circuit. The resistance or load in the system is what consumes/reduces energy. For example, a light bulb in an electric circuit is a load. Resistance determines how much current will flow through a component. Resistors are used to control voltage and current levels. A high resistance allows a small amount of current to flow. A low resistance allows a large amount of current to flow. Resistance is measured in ohms (represented by "R" or "Ω"). One ohm is the resistance value through which one volt will maintain a current of one ampere.

German physicist Georg Ohm

Ohm's Law

Ohm's Law was named after a German physicist, Georg Simon Ohm. Ohm's Law defines the relationships between voltage (V), current (I), and resistance (R). Ohm's law states that the amount of steady current through a material is directly proportional to the voltage across the material, for some fixed temperature. It takes one volt to push one amp of current through one ohm of resistance.

Ohm's Law is:

$V = I \times R$ where V = voltage (EMF)

The two variations of Ohm's law are:

$I = V/R$ where I = intensity of current (amperage)

$R = V/I$ where R = resistance (load)

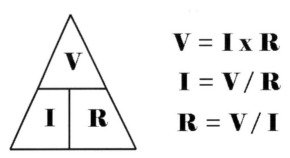

Ohm's Law defines relationships between voltage, current and resistance.

Other Measures of Power

James Watt improved the design of the steam engine, making its use practical. In order to communicate to people how powerful his steam engines were and to sell his steam engines, he needed to relate the power of steam engines to something people could understand. A foot-pound is the amount of force required to raise a one-pound weight to a height of one foot. In James Watt's experiments, he determined that an average horse working at a steady rate could do 550 foot-pounds of work per second.

1 hp = 550 ft-lbs/second

Further experiments determined that the amount of electrical energy required to do 1 horsepower was 746 watts.

1 hp = 746 watts

Another unit of energy is derived from the British system of measures. The British Thermal Unit (BTU) is the amount of heat required to raise the temperature of one pound of water one degree Fahrenheit. In the metric version, a calorie is the amount of heat needed to raise the temperature of one gram of water one degree Celsius.

The joule is defined as the amount of work done by one watt for one second.

Standard units of engineering notation

Engineering notation is used to simplify writing very large or very small numbers. For example, it is much easier to write 3 ns than to write 0.000,000,003 seconds or 15 M ohms instead of 15,000,000 ohms.

Standard units of engineering notation

Prefix	Symbol	Value	Description
pico	p	10^{-12}	0.000000000001
nano	n	10^{-9}	0.000000001
micro	m	10^{-6}	0.000001
milli		10^{-3}	0.001
centi	c	10^{-2}	0.01
deci	d	10^{-1}	0.1
kilo	k	10^{+3}	1,000
mega	M	10^{+6}	1,000,000
giga	G	10^{+9}	1,000,000,000
tera	T	10^{+12}	1,000,000,000,000

Concept Reinforcement

1. Explain the concept of an amp.

2. Explain resistance.

3. Explain voltage.

4. State Ohm's Law and the two derivations.

Section 1.4 – The Generation of Electricity

Section Objective

- Describe how electricity is generated

Electrical Current

Electricity is the flow of electrons through a path. The objective of electrical generation is to provide sufficient current (flow of electrons) through power lines to hospitals, police and fire stations, businesses, schools and homes to power the electrical devices used by people. This requires the generation of electrical energy from a power source. As you can imagine, it takes a lot of power to meet the demands of everyone. Where does energy come from in the first place? You have heard about coal burning power plants and nuclear power plants. You have also heard about alternative sources of power from renewable energy sources such as wind, solar, and water.

What is a Conductor?

Atoms with only 1, 2, or 3 valence electrons in their outer orbit give up their valence electrons when struck by another electron. The valence electrons absorb energy from a striking electron leaving their original orbit and moving on to strike other electrons. This movement of electrons creates a current that flows in the conductor. Conductors are used to carry electricity. A good example of an electric conductor is copper wiring since copper has only a single valence electron in its outer orbit.

What is a Generator?

A generator is a device that converts mechanical energy into electrical energy. A typical generator uses a large quantity of copper wiring wrapped around a shaft (called an armature) spinning inside very large magnets at very high speeds. Energy, or power, in the form of electricity comes from the generator and is transmitted through transmission lines (electrical power lines) and delivered to the end users. That includes you.

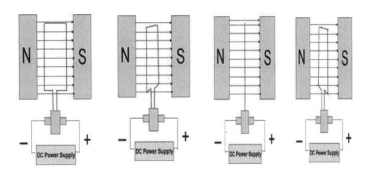

Diagram of a simple electric generator: as the armature (copper wiring) rotates between two magnets, current is produced in the copper wire. The battery is the power supply that rotates the armature.

To generate electricity, the armature must be turned by a force. In power plants, most often the armature is turned by the force of water or steam. Steam is generated by boilers.

A Little History

In 1831, a British Scientist named Michael Faraday discovered that when an electric conductor, such as a copper wire, is physically moved through a magnetic field, the magnetic field caused the electrons in the conductor to move. The movement of the electrons in the conductor created electric energy (electricity). This means that the mechanical energy that moves the conductor wire through the magnetic field is converted into electric energy that flows in the wire.

Michael Faraday

Types of Generators

Electricity can be generated in a number of ways. Batteries, coal-fired power plants, hydroelectric plants, and solar and wind energy are a few ways electricity is generated.

Batteries

Batteries generate electricity as a result of chemical reactions. Most batteries have two posts: one which is positive and one which is negative. Batteries do not generate any electricity unless the circuit is completed. The circuit is completed by connecting the terminals with a wire, which usually includes a load (a light bulb or some other item that requires energy to work).

Car Battery

Conversion of mechanical energy to electricity

Power plants generate electricity by converting the original source of energy to electricity through a mechanical process. Power plants use generators to convert mechanical energy to electricity. One component of a generator is a turbine. A turbine has a shaft with blades at one end and electromagnets at the other. Wind, water and steam are all used to push the turbine blades, which makes the turbine shaft and electromagnets spin very quickly. The ends of the electromagnets are surrounded with coils of copper wire. Remember that copper is a good conductor of electricity, so when the spinning magnets cause the electrons in the wire to begin to move, electricity is generated. Moving water moves the turbine blades that generate hydroelectric power.

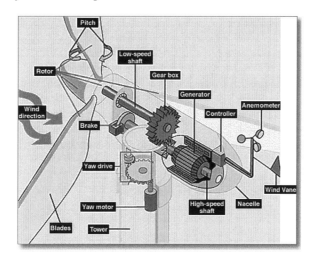

Schematic of the components of a wind turbine, which is used to generate electricity from wind.
Image courtesy of the US Office of Energy Efficiency and Renewable Energy

Concept Reinforcement

1. What is a conductor?

2. What does a generator do?

3. Name three sources of renewable energy that can be used to power a generator.

Section 1.5 – Power, Energy and Work

Section Objective

- Describe power and how it relates to energy and work

The Relationship Between Power, Energy and Work

Work is an activity involving force and movement in the direction of that force. Energy is the ability to do work. However, work and energy are essentially the same and are measured using identical units of measurements.

Work and Energy

Units of measurements for work and energy are joules, BTUs, and calories.

A **joule** is defined as the amount of energy it takes to lift or move an object that weighs one newton (N) a distance of 1 meter (m). One newton weighs 0.445 pounds, so a joule is the amount of energy it takes to move 0.445 pounds of something a distance of one meter. A joule is the energy required to pass 1 ampere of electrical current through 1 ohm of resistance in 1 second.

BTU stands for British thermal unit. A British thermal unit is equal to 1,055 joules. A BTU is also equal to 252 calories. A BTU is the amount of energy required to raise the temperature of one pound of water by one degree Fahrenheit. This is similar to the heat produced by burning a single wooden match.

A **calorie** is a unit that describes the amount of energy required to raise the temperature of one gram of water by 1 degree Celsius. A calorie is equal to 4.187 joules. One calorie is also equal to 0.03969 BTU. Another way to describe the calorie is the energy required to raise the temperature of one gram of water by one degree Celsius.

Work is power used over a period. Work is equal to the force needed to move an object multiplied by the distance the object is moved. The equation that shows this is work = force x distance, or $W = F \times D$.

Power is the rate at which work is performed or the rate energy is used. As an example, if you and your friend compete to see who can move 100 pounds over a distance of 10 yards the fastest, you are testing power. You are both doing the same amount of work, but the person who moves the 100 pound object over the 10 yard distance the most quickly shows more power.

Joules

The amount of energy required to move an object that weighs 1 newton (N) a distance of 1 meter (m).

1 Joule = 0.2388 calories.

1 Joule = 0.0009481 BTU

British Thermal Units

The amount of energy required to raise the temperature of one pound of water by one degree Fahrenheit.

1 BTU = 252 calories

1 BTU = 1,055 Joules

Calorie

The amount of energy required to raise the temperature of one gram of water by one degree Celsius.

1 cal = 4.187 Joules

1 cal = 0.03969 BTU

Power

Power is calculated using the following formula: power (P) = work (W) ÷ time (t)

There are two units of measurements for power. These are watts and horsepower (hp).

Wattage is the amount of power that is used in an electric circuit. Wattage is proportional to the amount of voltage and the amount of current flowing through an electric circuit. One watt is equal to the use (or production) of one joule of energy per second.

Horsepower is a unit of power measurement that is often associated with engines. One horsepower is equal to 746 watts.

Electric power is the rate at which electrical energy is transferred by an electric circuit. In order for power (watts) to exist, electrical energy must first be converted into heat or mechanical energy.

Conversion table for power units

Common power units	Equivalent value
1 horsepower	746 watts
1 horsepower	550 ft-lbs/second
1 watt	0.00134 horsepower
1 watt	3.412 BTU/hour
1 watt × second	1 joule
1 BTU/second	1.055 watts
1 calorie/second	4.19 watts
1 ft-lb/second	1.36 watts
1 BTU	1055 joules
1 joule	0.2388 calories
1 calorie	4.187 joules

Summary

Power is the rate at which work is performed. Electric power is the rate electrical energy is transferred by an electric circuit. Units of measurement are watts "W" and horsepower "hp." (1 horsepower = 746 watts) (1 watt = 0.0134 horsepower)

Work is power used over a period of time. Units of measurement are joules, BTUs and calories. (1 BTU = 1055 joules) (1 calorie = 4.187 joules)

Work and energy are basically the same and use the same units of measurement (joules, BTUs, and calories).

Concept Reinforcement

1. Explain the relationship between power and work.

2. List the units of measurement for work and energy.

3. List the units of measurement for power.

4. State the mathematical relationship between work and power.

Section 1.6 – What is an Electrical Circuit?

Section Objective

- Explain an electrical circuit

Electricity

Electrical current is the flow of electrons. When the valence electron of an atom is bombarded with sufficient energy to leave its current orbit, it will move to hit another atom. This movement of electrons is what creates a current and is commonly referred to as electricity.

Electrical Circuit

An electrical circuit is the path along which the electrons move. In other words, an electrical circuit is the path that electric current follows. Electrical current requires a power source. The electrons flow from the source along a path and return to the source. Just like a bicycle path, there is a beginning, a path along which the cyclist rides and then a return to the starting point.

There can be various things in the path that affect the flow of current. Like a bicycle riding uphill, the current may find a resistance that slows down the flow of electrons.

The cyclist may have to stop and carry his bike over an obstacle. Current may have to do some work before it can continue along its path. For example, if there is a lamp in the circuit, the current will convert some of its energy to light the lamp. Then the current continues past the lamp down the path back to the source of energy.

Cyclist following an irregular path

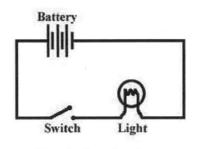

Circuit showing lamp

Definitions of Closed Circuits and Open Circuits

The circuit however large or small always returns to its source. This is called a **closed circuit**. If the circuit has a break in it, it is called an **open circuit** and no current will flow.

If the cyclist comes to a bridge and the bridge has fallen down, he cannot complete his path. He cannot complete his path unless he crosses the bridge. He stops at the fallen bridge and tries to figure a way to get across. Being a clever cyclist and always prepared, he has an axe in his backpack. He chops down some logs and fashions a bridge. With some difficulty, he is able to carry his bike across the makeshift bridge and continue down the path to complete his ride back to where he started.

An open circuit has a break in its path just like the cyclist found when he came upon the fallen bridge.

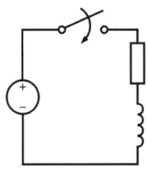

An open circuit

If something closes the circuit, like the cyclist completes his path with his makeshift bridge, then the electrons can continue to flow.

Definition of a Circuit Breaker

A circuit breaker is exactly what it sounds like. A **circuit breaker** breaks the circuit and changes a closed circuit to an open circuit. Circuit breakers provide safety in many situations. The places where you live, work and study have circuit breakers installed. There are codes the electricians must abide by when installing electrical cables. Circuit breakers monitor the flow of electrons through the electrical cables. If there is a surge of electrons too powerful for the circuit to handle, the circuit breaker overloads and trips. When it trips, it opens the circuit and stops the flow of current. If there was an overload and the circuit breaker did not trip, the heat created could damage an appliance or electronic device or cause a fire.

Circuit breaker

Have you ever plugged in too many devices in one socket and had all the devices shut down at once? Well, you overloaded the circuit with your demand to power too many devices at the same time. The circuit could not provide sufficient power to handle all of the devices at the same time. The circuit tried to supply the power, but the excess demand was too much for the circuit breaker. The circuit breaker tripped, opening the circuit and stopping the flow of all power to that circuit.

Did everything shut down? The answer should be no. If the electrician did his job correctly, he installed more than one circuit in the building. Each circuit should have its own circuit breaker.

If this has never happened to you, ask an older person. This was very common before electrical standards were created.

Definition of a Fuse

A **fuse** is similar in function to a circuit breaker. The major disadvantage of a fuse is that when it overloads, it "breaks" in order to create an open circuit. The fuse must then be replaced. A circuit breaker trips and can be reset once the problem that caused it to trip has been resolved.

Plug fuses are not made in ratings over 30 amp.

A typical Type-S non-tamperable fuse, and its adapter. Once an adapter has been screwed into a fuse-holder, it cannot be removed. This prevents use of fuses larger than originally intended.

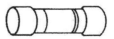
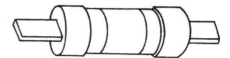

Cartridge fuses rated 60 amps. or less are of the ferrule type shown.

Cartridge fuses rated more than 60 amp. have knife-blade terminals shown.

Fuses are still used in many places. Your car probably has some fuses in the electrical system. When the fuse breaks, it must be replaced. Until it is replaced, the circuit will not work. Depending on the design of your car, you may find yourself without certain functions such as lights or heat.

Electric Meter

An electric meter is supplied by the local utility company and it measures the amount of current that flows into a home, school or business. The purpose of this meter is to be able to charge the consumer for the amount of energy used. If you were to turn off all the lights, your computer, TV, DVD player, video games, and all other electronic devices, you would still see the dials rotating on the meter. This means current is still flowing and being used. Even if you shut off the refrigerator and the furnace or air conditioner, you would still see the dials rotating. There is a lot of "hidden" consumption. As long as an appliance or device is plugged into the circuit, there is a flow of current. For example, if you look at your TV, it may be "powered" off, but if it is plugged in, there is probably a red light showing that the TV is still consuming power. It consumes less power when it is turned off, but it still uses a small amount as long as it is part of the closed circuit. A good way to save electricity is to carefully unplug devices not in use regularly. That doesn't mean you should go around and unplug everything. Some things need to stay plugged in, such as your refrigerator.

Electric Meter

Summary

An electrical circuit is the path along which electrons flow. A complete electrical circuit requires a power source and a closed path that returns to the source. It there is a break in the circuit so that the flow of electrons is interrupted, the circuit is an open circuit and no current will flow. When there is no break in the circuit, the circuit is closed and current will flow.

Circuit breakers and fuses provide a safety feature by opening the circuit when there is a power overload.

Concept Reinforcement

1. What is an open circuit?

2. What is a closed circuit?

3. What does a circuit breaker do?

4. What is the difference between a circuit breaker and a fuse?

Section 1.7 – Conductors, Insulators, Resistors, Variable Resistors and Capacitors

Section Objective

- Describe conductors, insulators, resistors, variable resistors and capacitors

Electric Circuit

When an energy source is applied to an atom, the valence electrons in the outer orbit are hit. All of the energy is applied equally to the orbiting valence electrons. If there are only a few valence electrons in the outer orbit, the energy will be sufficient to knock the electrons out of their current orbit, forcing them to move on to hit other electrons. If there are more valence electrons in the outer orbit, the energy will be dissipated among all of the valence electrons and there may be insufficient force to knock the electrons out of their current orbit. Electrons follow the path of least resistance. In other words, the electrons go to the atoms that are more unstable and more willing give up their electrons.

Conductor

The atoms with fewer electrons in their outer orbit are more unstable and will give up their electrons when hit. These atoms are called **conductors**. The electrons move along the conductive material and current flows easily through a conductor.

Example of a conductor

Insulator

When an atom holds on to its electrons, the material is considered an **insulator**. Current has a difficult time flowing through an insulation material. Electrons have a more difficult time moving through an insulator.

An example of how an insulator would be used would be to wrap material that is considered an insulator around material that is a conductor. An example is a copper wire (conductor) wrapped in rubber-like polymer (insulator). If the rubber-like polymer insulation were not wrapped around the copper wire, when electric current flowed through the wire, the electrons would follow any path of lower resistance. If you held the bare copper wire in your hand while current was flowing through it, OUCH! Some of the electrons would move into you and provide you with a nasty and perhaps lethal shock (depending on the amount of current).

Example of an insulator

Resistor

Resistor is somewhere between an insulator and a conductor. Current flows easily through a conductor and almost not at all through an insulator. Current does flow slowly through a resistor, however, it is like walking through mud. While you can walk through mud, it is difficult to do so and it slows you down. Current can flow through a resistor, but a **resistor** slows down the flow of current. Resistors dissipate energy in the form of heat.

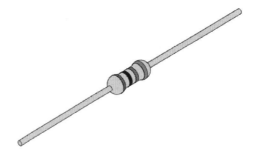

Example of a resistor

Variable Resistor

Variable resistor is exactly what it sounds like. It is a resistor that has the ability to change the amount of resistance it applies to the flow of current in an electrical circuit.

The variable resistor is on the front-left portion of the power supply.

Capacitor

A capacitor is a form of storage device for electrons. Capacitors store the potential energy of accumulated electrons in the form of an electric field.

Ceramic vacuum variable capacitor

Summary

The purpose of a conductor is to provide a path for electric current. A conductor is made of material that readily gives up its valence electrons providing a means for current to flow. An insulator protects the conductor. An insulator is made of material that does not readily give up its valence electrons making it difficult for current to flow through it.

A resistor is usually some form of load on the circuit. For example, a resistor might be a light bulb or an electric heater.

A capacitor is a storage device for electrons.

Conductors, resistors, and capacitors are among the "building blocks" used to create electronic devices.

Concept Reinforcement

1. What is a conductor?

2. What is an insulator?

3. What is a resistor? What is a variable resistor?

4. What is a capacitor?

Section 1.8 – Batteries and Their Uses

Section Objective

- Explain how a battery works

Definition of a Battery

A battery is a closed container full of chemicals that produce electrons. All batteries have a positive terminal and a negative terminal. When the battery is installed in a closed circuit, the chemical reaction of the chemicals in the battery generates electrons. The electrons generated by the chemical reaction provide a power source. Unless the terminals are connected through a circuit, there is no flow of electrons.

Close up of batteries showing + and – terminals.

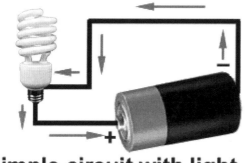

Simple circuit with light

Closed circuit with a battery as power source

A Little History

Allessandro Volta in 1800 repeated some experiments performed by Luigi Galvani in 1786 on frog legs. When Luigi touched the frog legs preserved in salt solution with a copper wire and an iron scalpel, the frog's legs twitched. Luigi thought the twitching of the frog's legs was caused by electricity. However, he thought the frog's contracting muscles had created the electricity. Allessandro Volta thought that it might be the chemical reaction of the salt, iron and copper that created the electricity. Further experiments by Allessandro Volta proved him to be correct. Volta created a voltaic pile which consisted of layers of zinc and silver separated by cardboard soaked in salt water. The resulting pile produced voltage and became the basis for the first practical battery.

Different batteries are made up of cells of varying materials. The amount of voltage produced by a battery is based on the materials and chemicals used in its construction.

Some Applications of Batteries

The list of applications is huge. As our society becomes more mobile, the need for more and more applications for batteries increases. From car batteries to cell phones, from flashlights to laptops, the one thing they have in common is a battery or multiple batteries. Batteries can be linked together either in series or in parallel. However, batteries of different voltages should not be connected in parallel.

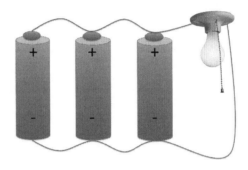

Batteries operating in parallel

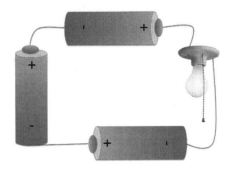

Batteries operating in series

Batteries provide a source of direct current in an electrical circuit.

Summary

A battery is a closed container full of chemicals. When a battery is connected in a closed circuit, electrons will flow through the circuit. Electric current is provided by the battery to power a device connected to the battery.

A battery is an independent power source providing direct electrical current.

Concept Reinforcement

1. What is a battery?

2. Explain a closed circuit.

3. List five items you use everyday that function with a battery.

Section 1.9 – The Difference between Alternating and Direct Current

Section Objective

- Explain the difference between alternating and direct current

Current

A conductive material or a conductor is made up of material that easily gives up its valence electrons when a charge is applied. Current is the movement of electrons in a conductive material when a charge is applied.

Direct Current

Thomas Edison developed a working central electricity generating station. This was the first commercial electric power transmission and it used direct current. Direct current is a closed circuit with the current flowing in one direction. The term **direct current** or **DC** is used to refer to power systems with unidirectional flow. The electrons always flow in a constant direction.

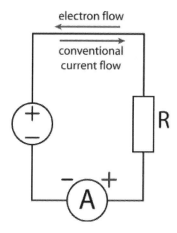

Direct Current Circuit Example

A battery produces direct current. Direct current is produced by other energy sources such as solar cells. Direct current is used to charge batteries. Direct current is used extensively to power adjustable-speed motor drives in industry and in transportation.

Alternating Current

Alternating current is a closed circuit with bidirectional current flow. The direction of the flow of electrons is reversed at regular intervals. The term **alternating current** or **AC** is used to refer to power systems that change polarity at regular intervals.

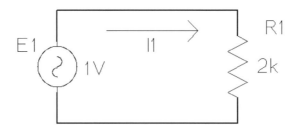

Alternating Current Circuit Example

Alternating current is the most common form of electrical power produced in the world. Alternating current is what powers the lights and appliances in your home. Some of the largest motors found in industry are powered by alternating current.

A **transformer** is a device that can step up or step down the voltage. The voltage is stepped up by a transformer for the purpose of transmission over long distances. At the destination, a transformer can step down the high voltage to a lower voltage so it can be safely used.

Why Use Alternating Current and Not Direct Current?

Thomas Edison developed the first commercially viable electric power when he built his electrical power distribution system providing power to 59 customers in lower Manhattan, NY. His power station provided direct current.

Nikola Tesla believed in alternating current. He joined forces with George Westinghouse to provide commercially viable alternating current to the consumer.

The Thomas Edison Company and the Westinghouse Company both submitted a bid to provide lighting for the 1904 World's Fair in St. Louis, Missouri. The Edison Company's bid was over a dollar per light while the Westinghouse Company's bid was under a quarter per light. Westinghouse Company used higher voltage alternating current. Edison Company used direct current.

Since Thomas Edison first produced commercially viable electric power and he produced direct current, it is curious to note that the bulk of power produced today is based on alternating current. Why the switch? It will not surprise you to find out the answer was an economic one.

A major advantage of alternating current is that it can be transformed from higher voltage to lower voltage while direct current cannot. Providing direct current over longer distances required larger cables. The larger cables were not only more expensive, but there were hazards involved in transmitting higher current over long transmission lines.

Transformers are used to step up voltage before transmitting alternating current over long lines. Transformers are also used to step down voltage at the end of the transmission. It is cheaper and easier to generate and to transmit alternating current over long distances.

The battle between Edison (direct current) and Tesla (alternating current) became known as the War of the Currents. While direct current is still used in many applications, the cheaper alternating current became the standard.

Converting Alternating Current to Direct Current

Direct current is used to power batteries and a number of other devices. A **rectifier** is an electrical device that converts alternating current to direct current.

Summary

Electricity is transmitted as alternating current over long distances. Transformers are used to step up alternating current voltage for transmission. Transformers are then used at the end of the transmission line to step down the voltage for usage by consumers.

Direct current is current that always flows in one direction. Batteries are a source of direct current. In alternating current, the direction of the flow of electrons is reversed at regular intervals. Normal household current in most countries is alternating current.

Concept Reinforcement

1. What is direct current?

2. What is alternating current?

3. Explain the function of a transformer.

Section 1.10 – Definition of Static Electricity

Section Objective

- Describe static electricity

What is Static Electricity?

Current occurs when electrons move. Static means stationary or non-movement. So if electricity is based on the movement of electrons and static means no movement, just what is static electricity? Sounds contradictory, doesn't it?

You have probably experienced the results of static electricity many times:

- When you walked across a carpet and turned a metal knob on a door.
- When you touched a light switch and felt a sharp tingle.
- When your hair stood on end after you removed your wool cap.

Atoms are made up of neutrons, protons and electrons. Neutrons have no charge. Protons have a positive charge and electrons have a negative charge. If the atom has an equal number of protons and electrons, the overall charge of the atom is neutral. What would happen if something removes one or more of the valence electrons orbiting around the nucleus of the atom? If the atom loses some of its valence electrons and these electrons are not replaced with other electrons, the overall charge of the atom will be positive. This is because the atom has more protons (which are positive) than electrons (which are negative).

On the other hand, if the atom acquires new valence electrons without losing any of its original valence electrons, the overall charge of the atom will be negative. This is because it has more negative electrons than positive protons.

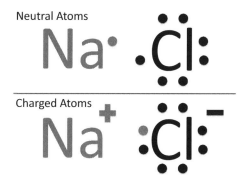

Atom losing electrons

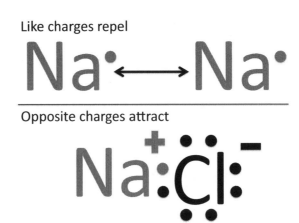

When electrons are lost or gained, the result is an imbalance. Now the atom has a negative or a positive charge. This imbalance of charges is known as **static electricity**. Like charges repel and opposite charges attract.

Charging an Object

How does an atom lose or gain electrons? It doesn't happen by itself. Something has to happen to the atom.

If you rub a piece of glass with a piece of wool cloth, you will find that the wool cloth sticks to the glass. In other words, the wool cloth and piece of glass are attracted to each other.

If you remove a wool hat from your head, your hair will probably "stand on end." Your hairs are repelling each other.

Try sliding your feet as you walk across a carpet and then touch a metal doorknob. Ouch!

So what happened? The valence electrons in the outer orbit moved from one material to another.

If you rub a piece of glass with a piece of wool cloth, electrons will move from the glass to the wool. The glass will then be positively charged because it has more protons than electrons. The wool, which has more electrons than protons, will be negatively charged. Since opposites attract, the wool will be attracted to the glass.

So why do your hairs stand on end when you drag a wool cap off your head? When you rub the wool hat against your hair, electrons move from your hair to the hat. The hairs become positively charged because they lose some electrons. Like charges repel. Each hair will be positively charged and will be repelled by the other positively charged hairs. The hairs will try to move away from each other and thus will look like they are standing up on end.

Discharging an Object

As you walk across the carpet, electrons move from the rug to you. You then have a negative static charge. The metal door knob is a conductor. When you touch the door knob, the electrons will jump from you to the knob. When they leave you and move into the knob, you will probably feel a shock. This is a discharge of electrons from you to the doorknob. You no longer have a negative static charge. However, if you slide your feet on the carpet again, you can build up another static charge.

Once a static charge builds up in an object, it can be discharged to another object by coming into contact with that object. The discharge of the electrons from a negatively charged object can be fast if the second object is a conductor, such as the metal doorknob. If the static buildup is large enough, you will feel the rapid discharge of electrons from you to the doorknob in the form of a minor shock.

Summary

Static electricity is the result of movement of electrons from one object to another creating a negative or positive charge. This imbalance of charges is known as **static electricity**. Like charges repel and opposite charges attract.

Concept Reinforcement

1. What does the word static mean?

2. Explain static electricity.

3. Explain the discharge of static electricity.

Section 1.11 – The Light Bulb

Section Objective

- Explain how a light bulb works

Electrical Circuit

An electrical circuit is a path along which electrons flow. The flow of electrons creates an electrical current along the circuit. A circuit must have a power source and a return path to the source. In order for the electrons to flow, the circuit must be a complete path with no breaks or interruptions in the flow of electrons.

Electrical Cable

An electrical cable provides a path for the flow of electrons. An electrical cable is made up of a conductor protected by an insulator. A conductor is made up of atoms that will give up the electrons (valence electrons) in their outer orbit when a charge or power is supplied at the source. A good example of a conductor is a copper wire. If you have a copper wire and apply a charge to one end, electrons will flow along the length of the wire. However, to insure that the electrons flow along the copper wire (desired path), the wire should be wrapped in a material that holds onto its electrons tightly and does not allow the copper valence electrons to replace its valence electrons. The material used to wrap the wire should be an insulator.

When a power source is connected to one end of the cable, the electrons flow along the conductive portion of the cable.

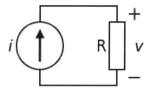

Drawing of a closed circuit (power source, path, return to source)

If the circuit is interrupted or broken, then the electrons will not flow.

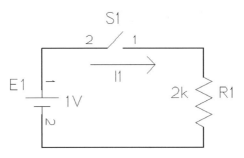

Drawing of an open circuit (power source, path, break in path, return to source)

A Little History

In the late 1700s and early 1800s, a large number of inventors were trying to create a means by which electric current would provide light. Thomas Edison is credited with inventing the first practical electric light. Thomas Edison had his own glass blowing operation and a number of assistants to create the bulbs he used for testing.

Thomas Edison tested thousands of materials including numerous plant fibers as well as metals in his search for a filament that would support the flow of electrons

He built his first high resistance, incandescent electric light by passing electricity through a thin platinum filament in a glass vacuum bulb. The vacuum in the bulb delayed the filament from melting from the heat given off by the electric current. The lamp only burned for a few short hours. Further testing by Edison and his team led to the use of carbonized cotton thread filaments. This provided a light equivalent to a 16 watt bulb and lasted for 1500 hours.

Today, the filament in an incandescent light bulb is usually made of a long, thin length of tungsten metal. For a 60-watt bulb, the tungsten filament would be about 6.5 feet long but only 0.001 inch thick. The metal is coiled to fit inside the bulb.

How Does a Light Bulb Work?

Two metal wires go up into a glass bulb. Inside the bulb, the two wires are connected by a filament. The glass bulb is filled with an inert gas.

> Thomas Edison has been credited with inventing the electric light bulb. However, he was not the first. Joseph Swan, a British inventor, working at the same time as Edison, obtained the first patent for the same light bulb in Britain one year prior to Edison's patent date.

Each wire from the bulb is connected to the wire in the electrical circuit. A power supply is connected to the circuit. Electric current flows up one wire through the filament and back down the other wire, completing the circuit.

When power is supplied to the circuit, the electrons move along the wire and through the filament. The electrons bump into the atoms in the filament causing movement of the atoms in the filament. The electrons in the filament atoms are not readily given up like in the wire (which is a conductor), but they are agitated sufficiently to orbit in a higher orbiting level around their nucleus. However, they are each still bound to its nucleus. The pull of the nucleus causes the electrons to fall back into their normal orbit. When they fall back into their orbit, they release the extra energy in the form of photons. These photons provide light.

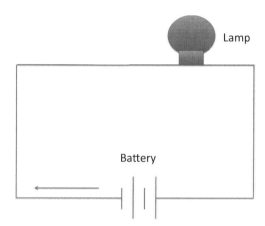

You want to be able to turn off the light when it is not needed. To turn off the light, you only need to open or break the circuit. A light switch introduced into the electrical circuit will provide the means of shutting off the light.

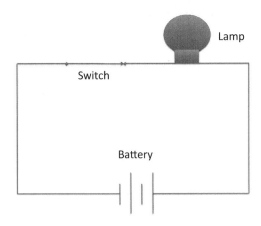

Drawing of a closed circuit with a light, power source and switch

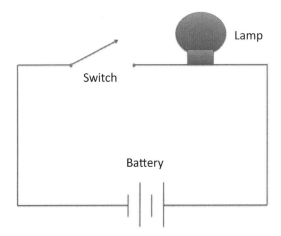

Drawing of a open circuit with a light, power source and switch

Disconnecting the circuit from its power source would also stop the flow of electrons and turn off the light.

Lamp

If you look at the inside of the light bulb, you should be able to see a fine wire connecting the two metal wires coming up from the base of the bulb. If the fine wire is broken the circuit cannot be closed, and the bulb will not light. The bulb is no longer functional and should be disposed of according to local regulations.

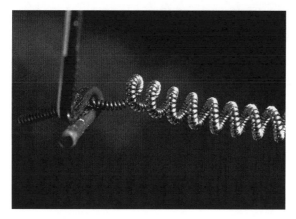

Highly magnified photo of a 200 watt light bulb filament.

Portable light can be obtained by using a battery as a power source.

Flashlight

Summary

When a light bulb is part of a closed circuit and power is supplied to the circuit, electrons will flow through the wire in the light bulb. The electrons move along the wire inside the light bulb and through the filament. The electrons bump into the atoms in the filament causing movement of the atoms in the filament. The electrons in the filament atoms are agitated but provide sufficient resistance to the electrons that they will fall back into their normal orbit. When they fall back into their orbit, they release the extra energy in the form of photons. These photons provide light.

Concept Reinforcement

1. When a light is on, is the circuit open or closed?

2. When a lamp is unplugged, is the circuit open or closed?

3. When a lamp is plugged into an electric socket and turned off, is the circuit open or closed?

Section 1.12 – Magnets and Magnetic Fields

Section Objective

- Describe magnets and magnetic fields

Magnets

Put two magnets together. One side of a magnet attracts one side of another magnet. Switch the magnets about and they repel each other. Why?

Magnetic Field

To understand magnets, it is necessary to understand atoms. The nucleus of the atom is comprised of tightly bound protons and neutrons. The negative electrons orbit around the nucleus of an atom. The electrons also spin as they orbit around their nucleus. Generally valence electrons spin as pairs in opposite directions while their electrons continue to orbit around their respective nucleus. By spinning in opposite directions, these electrons will cancel out the negative effect of each other.

However, if the valence electrons in adjacent atoms spin in the same direction, this is what is believed to create a magnetic field. Their electrons combine to form a spin pattern which causes the magnetic fields to add rather than cancel each other out. A magnetic field is the space in which a magnetic force exists.

> Since the North Pole acts like one end of a giant magnet and the South Pole acts like the other end, the magnetized freely balanced needle will rotate from anywhere on the earth to point to one end of the Pole and away from the other end. This is very useful in navigation.

Types of Magnets

Magnets can be found in nature, but they can also be created. Some materials can be magnetized easily and some cannot. Some magnets are permanent magnets and some lose their magnetism.

Magnetite and lodestone are two magnets that occur naturally. An iron bar can be made magnetic by rubbing it with magnetite, lodestone, or other magnetic material.

An **electromagnet** is usually made from a coil of conductive wire. When an electric current passes through the wire, it acts like a magnet. When the current stops, it loses its magnetic capability.

Why Magnets Are Important to Electricity

When an electric conductor, such as a copper wire, is physically moved through a magnetic field, the magnetic field causes the electrons in the conductor to move. The movement of the electrons in the conductor is electric current.

This means that the mechanical energy that moves the conductor wire through the magnetic field is converted into electric energy that flows in the wire.

A generator is the device that converts the mechanical energy into electrical energy. A typical generator uses a large quantity of copper wiring wrapped around a shaft (called an armature) spinning inside very large magnets at very high speeds. Energy or power in the form of electricity comes from the generator and is transmitted through transmission lines (electrical power lines).

Summary

A magnet produces a magnetic field around it. Some magnets are found naturally in nature and others are created.

When current is passed through a coil of conductive material, the coil behaves like a magnet. The magnetic properties of the coil disappear when the current is stopped. This is an electromagnet. A large coil of conductive wiring (usually copper) wrapped around a shaft is called an armature.

Magnets are important to the generation of electricity. A generator uses a spinning armature inside large magnets to convert mechanical energy into electrical energy. Mechanical energy turns the armature inside the magnets creating current.

Concept Reinforcement

1. What is a magnet?

2. Name two magnets found in nature.

3. What is an electromagnet?

Section 1.13 – The Generation of Heat by Electricity

Section Objective

- Describe how heat is generated by electricity

Electrical Circuit

An electrical circuit consists of a source of electrons, a device that uses the electrical current and a path along which the electrons will flow. Now consider obstacles to the flow of electrons. What happens if there is resistance to the flow of electrons?

Moving electrons have energy. As the electrons move from one point to another, they can do work. In an incandescent light bulb the energy of the electrons meets resistance when going through the filament. The energy is converted to heat and the heat in turn creates light. Circuits may have one resistor or many resistors.

When electrons meet resistance, energy is dissipated into heat and this heat is lost to the circuit.

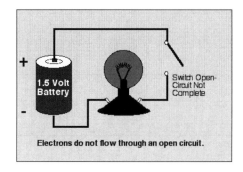

Electrons do not flow through an open circuit.

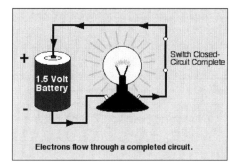

Electrons flow through a completed circuit.

The energy that is lost to heat is not returned to the source and is lost to the circuit. The circuit must remain closed in order for current to continue to flow. Additional energy is drawn from the source in order to continue to overcome the resistance in the closed circuit.

Ohm's Law

In a direct current circuit, the amount of current (flow of electrons) is directly proportional to the voltage supplied (power source) and inversely proportional to the resistance. Ohm's Law states that it takes one volt to push one amp of current through one ohm of resistance.

Summary

Moving electrons have energy. As the electrons move from one point to another, they can do work. When electrons meet resistance, energy is dissipated into heat and this heat is lost to the circuit. Additional energy must be drawn from the energy source in order to replace the heat lost if the circuit is to remain in the same state.

Concept Reinforcement

1. What causes heat in an electric circuit?

2. State Ohm's Law.

3. Explain how a light bulb helps create heat.

Section 1.14 – Grounding

Section Objective

- Describe grounding and explain its importance

Gremlins

Systems that are improperly grounded are likely to have problems. Computer control systems can act "funny" and strange things can happen. Sometimes computer systems behave like they have been invaded by gremlins. Unexplained glitches and program interruptions can occur. Programmers and Systems Engineers work hard to figure out what is going on. Sometimes it is as simple as the current that provides power to the systems being corrupt due to grounding faults.

On a more serious note, if a building is not properly grounded, the results can be quite disastrous. A proper earth ground safely conducts stray electrical current to earth for personal safety. For example, a telephone installation person can be seriously shocked if the building is not properly grounded.

Lightning striking a building that is not grounded could damage electrical equipment or even cause a fire.

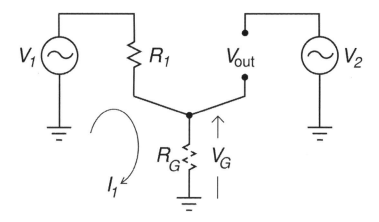

Circuit diagram illustrating a ground loop

Two circuits share a common ground wire. The voltage of the ground connection, VG, should ideally be zero. However if the ground conductor has significant resistance, RG, the current flowing through it from circuit 1 will cause a voltage drop since VG = I₁RG. Since they share the ground wire, this voltage appears as unwanted interference in the second circuit: Vout = V2 + VG.

Grounding

Lack of proper grounding sounds pretty ominous, doesn't it? So just what is grounding?

The purpose of **grounding** is to provide a drainage system for excess buildup of electrical charge. A ground connection is a direct link to the earth. Good old earth can absorb or dissipate an unlimited amount of electrical charge. So when a buildup of electrical charge occurs, such as lightning or excess static electricity, the ground connection provides the path to discharge that excess charge.

When properly grounded, a point in the system tends to remain at a constant voltage no matter what happens elsewhere in the system.

When properly grounded, excess electrical charge will be dissipated into the ground and will not wander about the electrical circuit looking for a means to discharge itself. That exit could be an unsuspecting person who touches an electrical device in the ungrounded circuit. An ungrounded circuit could result in the inner workings of a computer being "fried" Erratic behavior in electrical devices can also be the result of improper grounding.

Summary

Grounding provides a safe outlet for the excess buildup of electrical charge. Ground is a direct link to the earth.

Concept Reinforcement

1. What is the purpose of grounding?

2. If a circuit is not properly grounded, is it dangerous?

3. If you are outside and lightning strikes do you want to be under a tree or lying flat in a ditch?

4. Do you want to play golf during a thunderstorm?

Section 1.15 – Circuit Breakers

Section Objective

- Describe a circuit breaker and explain how circuit breakers save lives

Closed Circuit

A closed circuit is an electrical path along which electrons flow with no breaks or interruptions in the flow of electrons. A circuit must have a power source and a return path to the source.

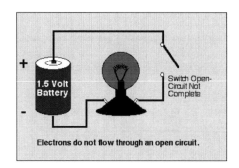

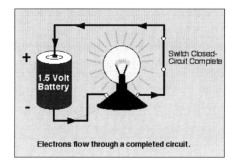

Example of open and closed circuits

Open Circuit

An open circuit is an electrical path that has some form of interruption to the flow of electrons. The circuit is broken at some point between the source and the return to the source.

Circuit Breaker

A circuit breaker is a safety device that will interrupt the flow of electrons when necessary. The sole purpose of the circuit breaker is to interrupt the power in a circuit.

A circuit breaker is designed to open (stop the flow of current) under an abnormal amount of current flow. If the flow of current is too much for the design of the circuit, the circuit is overloaded. The circuit breaker trips (opens) and the flow of electrons stops. Anything that is powered by that circuit will no longer work. If the circuit breaker did not trip under the excess load, the danger of fire due to overheating is eminent.

Circuit breakers are safety devices. They are rated to handle a certain amount of current and as long as the current is within the permissible limits, current will flow and power will be supplied to all electrical devices connected to this particular circuit.

> Before circuit breakers were used, fuses were installed in electrical panels to provide protection against power surges. When a surge of electrons occurred, the fuse would melt creating an open circuit. If there were no replacement fuses available, sometimes people would replace the fuse with a copper penny. While this restored power, it was extremely dangerous because the penny would not provide protection against future power surges. Remember, there was a problem to begin with because the fuse blew. The problem may not have been solved.

Panel of circuit breakers. Note that the plug to the left should be a ground fault circuit interrupt (GFCI) plug instead of a standard grounded plug.

In the photo are a number of switches. Each one is a circuit breaker for an electrical path. Each electrical path connected to a circuit breaker is independent from the other electrical paths. That is why if one circuit breaker trips, it does not affect the other electrical paths. So some lights and electrical devices may work, while the ones connected to the circuit breaker that tripped do not work.

There is a master circuit breaker which if tripped, will shut off all flow of electrons or power to everything connected to the electrical panel. It would take a major surge of electrons to trip the master circuit breaker. If this happened, all power would be shut off. If a major surge occurred and the breaker did not trip, worst case would be a fire. Less serious would be blown electrical equipment. That is why most computer manufacturers require surge protectors on the electrical strips where you plug in your computer equipment. A surge of electrons, if large enough, can "fry" the electrical components in your electronic devices. That means the internal components in your equipment overheat causing a meltdown. Static electricity, which is the build up of electrons, when discharged can cause a surge.

Electric current can be dangerous. Circuit breakers are a safety device. The circuit breaker also provides a means to easily shut down power so that someone can work on the wiring safely. If someone installs new wiring for an electrical appliance such as a stove, it is necessary to power down the circuit.

Summary

Circuit breakers are designed to trip (open the circuit) if there is too much current flowing through it. This is a safety feature. A surge could be caused by lightning, or an overload of electrical appliances on that circuit.

Circuit breakers are also used when someone is working on the wiring. By manually tripping the circuit breaker, current through that circuit is stopped. Now someone can safely do repairs or maintenance on the wiring.

Concept Reinforcement

1. What is an open circuit?

2. What is a closed circuit?

3. What is a circuit breaker?

4. When should you open a circuit breaker manually?

Unit Two

Section 2.1 – Electronic Symbols   70

Section 2.2 – Ohm's Law   75

Section 2.3 – Basic Circuit Concepts – Series Circuits   79

Section 2.4 – Basic Circuit Concepts – Parallel Circuits   83

Section 2.5 – Basic Circuit Concepts – Combination Circuits   85

Section 2.6 – Basic Circuit Concepts – Kirchhoff's Current Law   87

Section 2.7 – Basic Circuit Concepts – Kirchhoff's Voltage Law   91

Section 2.8 – Basic Circuit Concepts – Solving Circuits   95

Section 2.9 – Basic Circuit Concepts – Thévenin's Theorem   101

Section 2.10 – Basic Circuits Concepts–Norton's Theorem   105

Section 2.11 – Basic Circuit Concepts – Superposition Theorem   107

Section 2.12 – Basic Circuit Concepts – Millman's Theorem   115

Section 2.13 – Measuring Instruments   119

Section 2.14 – The Wheatstone Bridge   123

Section 2.15 – An In-Depth Review of Circuit Concepts   125

Section 2.1 – Electronic Symbols

Section Objective

- Describe electronic symbols and explain their importance

The Purpose of Electronic Symbols

What do Egyptian Hieroglyphics, Roman Numerals, the Greek alphabet, Arabic numbers, the English alphabet, C++ and electronic symbols have in common? They all provide a means to communicate ideas or concepts in writing. Words and symbols in any language provide the means to put an idea or thought in writing so that someone else can read and understand the message being communicated.

Electronic symbols are the written language of circuits. An electric circuit is a path from a source of power through various obstacles and devices with a return to the source. The components included in a circuit depends on its purpose or application. The circuit could be a hard wired loop that lights a battery powered light. A circuit could also be one of the many that make up the electronics in your car.

So if you are going to build something based on existing electronics, wouldn't it be nice to be able to read the electrical drawings and understand what is going on? What if you had a great idea and wanted to communicate that idea to the patent lawyers and to the manufacturers that would build your product? A common language for communicating electrical design is necessary.

Electrical drawings are called electrical **schematics**. Electrical schematics are used to build electrical equipment. Electrical schematics are used by builders to build homes and buildings. Electrical schematics are actually plans for electric circuits. So you can see that having a consistent method to represent electrical circuits is very important.

In the image below is a sample electrical schematic. It is very simple showing only a source and a light.

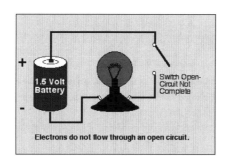

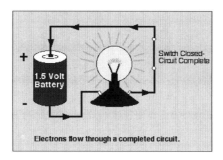

Sample of a very simple electrical schematic
Image courtesy of Fermi National Accelerator Laboratory Education Office

Here is a sample of a more complex electrical schematic. Can you imagine trying to explain what is going on in the circuit without being able to look at a document written using a standard set of symbols?

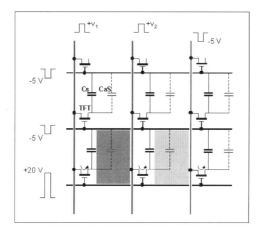

Sample of a more complex electrical schematic

Electronic Symbols

Electronic symbols are the building blocks of electrical circuits. In this section you are going to learn a few of the basic building blocks of electrical circuits. Each one of these symbols represents a function.

Power Sources

First you will look at power sources. In this session you will learn two different symbols for power. The one shown below is the symbol for a battery.

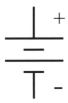

Battery (power source)

Power can also be obtained from a generator. The symbol below is the one used to represent a generator in an electrical schematic.

Generator (power source)

Flow of Current

In any drawing or schematic, it is important to show the flow of the current. In an electrical schematic, the path for the flow of current is shown as a solid line. The solid line in a schematic represents a conductor which transmits current. This is the path along which the electrons will travel when the circuit has power applied and is closed.

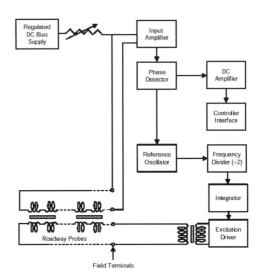

Example of an electrical path

Resistors

Current can flow through a resistor, but the purpose of a **resistor** is to slow down the flow of current. Resistance determines how much current will flow through a component. Resistors are used to control voltage and current levels. A high resistance allows a small amount of current to flow. A low resistance allows a large amount of current to flow. The symbol for a resistor is shown below.

Resistor symbol

Sometimes it is important in a circuit to be able to adjust the resistance. That is when a **variable resistor** is used.

Variable Resistor Symbol

Capacitors

A **capacitor** is a form of storage device for electrons. Capacitors store the potential energy of accumulated electrons in the form of an electric field.

Capacitor symbol

Summary

The symbols in this session are a few of the basic building blocks used to represent the contents of an electrical circuit. The graphic representation of these symbols in the form of an electrical diagram are called electrical schematics and are used to convey ideas, document designs and provide the basis for building anything electronic.

Concept Reinforcement

1. What is the purpose of electronic symbols?

2. What is an electrical schematic?

3. Draw the symbols for batteries, generators, resistors and capacitors.

4. How would you represent the flow of current in a schematic?

Section 2.2 – Ohm's Law

Section Objective

- Explain Ohm's Law

Definitions

Electrical Circuit

Electrons flow through a path. The complete path including a return to the source is an electric circuit.

Coulombs – a standard unit representing electric charge

A coulomb is a unit representing electric charge. One coulomb equals 6.25×10^{18} electrons. One coulomb is equal to one amp of current flowing through a conductor for one second.

Amps – a standard unit of electric current

The ampere (amp) is a measurement of the amount of electricity that flows through a circuit. An amp or ampere is represented by "I" in a formula. One ampere of current is defined when one coulomb (electrons) passes a point in a wire in one second.

Joule – a standard unit of work

The joule is defined as the amount of work done by one watt for one second.

Joule = watt × second

Volt – a standard unit of energy

Voltage is the force that pushes electrons through a wire. Voltage is also defined as electromotive force (EMF). You can think of voltage as the electrical pressure which pushes electrons through the electric circuit. Voltage does not flow through a wire, it pushes current (electrons) through a wire. A volt causes one coulomb to produce one joule of work. Voltage drop is the difference in voltage from one end of an electrical circuit to the other. Voltage is the difference in electrical potential between two points in a circuit and is measured in volts. Volts are represented by "V."

$V = I \times R$

Watt – a standard unit of power

One watt is one ampere of current flowing at one volt. Wattage is the amount of power that is used in an electric circuit. The unit of measurement is the watt and is represented by "W" (sometimes you will see "P"). Electric power is the rate electrical energy is transferred by an electric circuit. Power in watts = voltage times current:

W = V × I

Ohm – a standard unit of resistance (load)

An ohm is the amount of resistance that allows 1 amp of current to flow when 1 volt is applied to the circuit. The resistance or load in the system is what consumes/reduces energy. Resistance is measured in ohms and is represented by "R" (sometimes you will see the Greek symbol Omega "Ω").

R = V/I or   Ω=V/I

Ohm's Law

Ohm's Law was named after a German physicist, Georg Simon Ohm. Ohm's Law defines the relationships between (V) voltage, (I) current, and (R) resistance. Ohm's law states that the amount of steady current through a material is directly proportional to the voltage across the material, for some fixed temperature. It takes one volt to push one amp of current through one ohm of resistance.

Ohm's law is:

V = I × R where V = voltage (EMF)

Two variations of Ohm's Law are:

I = V/R where I = intensity of current (amperage)

R = V/I where R = resistance (load)

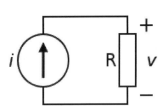 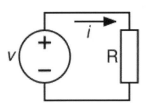

This diagram shows the current in and voltage across are resistor being driven by an independent current source as functions of time (rather than phasors, as a capital letter suggests). It uses the modern symbol for a resistor.

This diagram shows the current in and voltage across are resistor being driven by an independent voltage source as functions of time (rather than phasors, as a capital letter suggests). It uses the modern symbol for a resistor.

The benefit of Ohm's Law is that it provides a relationship between voltage, current and resistance. Ohm's Triangle is a useful tool for remembering these relationships.

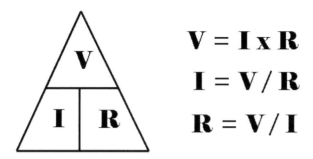

$$V = I \times R$$
$$I = V / R$$
$$R = V / I$$

Ohm's Law Triangle. This is a tool to help you remember Ohm's Law.

Summary

Electricity is a complex topic. Additionally, because it is not typically visible, electrical engineers have developed a series of symbols to describe the various components of an electrical circuit, as well as the relationships that develop when electricity interacts with physical objects. There are several units of measure that describe electricity in different ways. Coulombs are standard units of electric charge. Amperes (Amps) are the standard unit of electrical current. The Joule is a standard unit of work, the volt is a standard unit of energy, the watt is a standard unit of power and the Ohm is a standard unit of resistance. Ohm's law describes the relationships between voltage, resistance and current. Electrical engineers must understand Ohm's Law.

Concept Reinforcement

1. What is the standard unit of electric current?

2. What is a standard unit of power?

3. What is a standard unit of resistance?

4. What is Ohm's Law?

Section 2.3 – Basic Circuit Concepts – Series Circuits

Section Objective

- Describe series circuits

A Review of Ohm's Law

Ohm's Law defines the relationship between voltage, current and resistance.

$V = I \times R$ where V = voltage (EMF)

$I = V/R$ where I = intensity of current (amperage)

$R = V/I$ where R = resistance (load)

Definition of Electrical Circuit

An electric circuit is the complete path along which electrons flow, including a return to the source. Electrical circuits are used in all electrical devices.

Definition of a series circuit

A **series circuit** is a circuit where current only flows along one path. Current is the flow of electrons. Because there is only one path for current to flow through, **the value of the current is the same at any point in the circuit.**

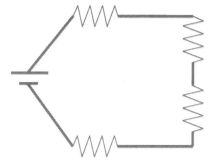

Series circuit diagram showing four resistors and a battery

Notice that there are four resistors in the circuit. **Voltage** is the force that pushes the electrons across the resistors. As the current passes through a resistor, the voltage will drop depending on the physical properties of the resistor. The amount of voltage dropped over the resistor can be measured by a voltmeter. **In a series circuit, the amount of voltage drops across all the resistors must equal the voltage applied to the circuit.** The total resistance is the sum of the individual resistors. To find out the total resistance of the series circuit, you need only add the values of all of the resistors in the circuit.

$$R_{total} = R_1 + R_2 + R_3 + R_4$$

$$R_{total} = 20 + 10 + 16 + 2$$

$$R_{total} = 48 \text{ ohms}$$

The voltage applied to the series circuit must equal the sum of the voltage drops across all the resistors in the circuit.

$$V = I \times R_{total}$$

Remember that in a series circuit, the value of the current is the same at any point in the circuit.

Knowing the value of the total resistance and the voltage applied to the series circuit, using Ohm's Law, you can calculate the current (I).

$$V_{total} = 96 \text{ volts and } R_{total} = 48 \text{ ohms}$$

$$I = V/R$$

$$I = 96 \text{ volts}/48 \text{ ohms}$$

$$I = 2 \text{ amps}$$

The current is the same flowing through each resistor.

$$I_{total} = I_1 = I_2 = I_3 = I_4$$

Using Ohm's Law and knowing the current is the same across each resistor, the voltage drop across each resistor can be calculated.

$$V_1 = I_1 \times R_1$$

$$V_1 = 2 \text{ amps} \times 20 \text{ ohms}$$

$$V_1 = 40 \text{ volts}$$

It takes 40 volts to push 2 amps of current through a 20 ohm resistor.

Summary

In a series circuit:

- There is only one path for current to flow.
- Current is the same at any point in the circuit.
- The total resistance is the sum of the individual resistors.
- The amount of voltage drops across all the resistors must equal the voltage applied to the circuit.

Concept Reinforcement

1. Give the three forms of Ohm's Law.

2. What are the four rules that apply to a series circuit?

3. Using the values given in the instruction above, calculate V_2, V_3 and V_4.

4. Add the values of V_2, V_3 and V_4 to V_1. Do the values add up to V_{total}?

Section 2.4 – Basic Circuit Concepts – Parallel Circuits

Section Objective

- Explain Kirchhoff's Law

A Review of Ohm's Law

Ohm's Law defines the relationship between voltage, current and resistance.

$V = I \times R$ where V = voltage (EMF)

$I = V/R$ where I = intensity of current (amperage)

$R = V/I$ where R = resistance (load)

Definition of electrical circuit

An electric circuit is the complete path along which electrons flow, including a return to the source. Electrical circuits can be organized as series, parallel or combination circuits.

Definition of a Parallel Circuit

A **parallel circuit** is a circuit where current flows along more than one path. Current is the flow of electrons. Because the current can flow through different paths, **the value of the voltage across any part of the circuit must be the same as the total applied voltage.** This is known as **Kirchhoff's Voltage Law**.

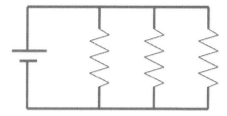

Parallel circuit showing three parallel paths

In a parallel circuit, the **total current** is the sum of the currents through all of the branches.

$I_{total} = I_1 + I_2 + I_3$

The fact that the amount of current that enters a junction must be equal the amount that leaves the junction is known as **Kirchhoff's Current Law**.

In a parallel circuit, the total resistance is the reciprocal of the sum of the reciprocals of the individual branches.

$1/R_{total} = 1/R_1 + 1/R_2 + 1/R_3$

Ohm's Law applies to the calculations of the circuit.

Summary

In a parallel circuit:

- There are multiple paths for current to flow.
- The total current is the sum of the current that flows through all the branches of the circuit.
- The voltage across any part of the circuit is the same as the applied voltage.
- The total resistance is the reciprocal of the sum of the reciprocals of the resistance of each branch of the circuit.

Concept Reinforcement

1. Give the three forms of Ohm's Law
2. What are the four rules that apply to a parallel circuit?
3. In the sample circuit shown, what is the applied voltage across the parallel circuit?

Section 2.5 – Basic Circuit Concepts – Combination Circuits

Section Objective

- Describe combination circuits

Definition of Electrical Circuit

An electric circuit is the complete path along which electrons flow, including a return to the source. A series circuit consists of a power source with resistors lined up in a series. The power follows only one path in a series circuit. Parallel circuits have several resistors lined up in parallel so the power can flow in multiple paths.

Definition of a Combination Circuit

A **combination circuit** is a circuit that contains both series circuits and parallel circuits. Current is the flow of electrons. A combination circuit operates like a series circuit when the wiring is in series. It operates like a parallel circuit when the electrons reach a junction point in the circuit where it has multiple paths to follow. A junction point where the circuit branches out from a series circuit into a parallel circuit is called a **node.**

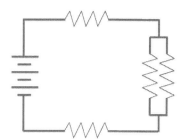

Combination circuit with series circuit and parallel circuit

Rules for Combination Circuits

For the series portion of the combination circuit

- There is only one path for current to flow.
- Current is the same at any point in the circuit.
- The total resistance is the sum of the individual resistors.
- The amount of voltage drops across all the resistors must equal the voltage applied to the circuit.

For the parallel portion of the combination circuit

- There are multiple paths for current to flow.
- The total current is the sum of the current that flows through all the branches of the circuit.
- The voltage across any part of the circuit is the same as the applied voltage.
- The total resistance is the reciprocal of the sum of the reciprocals of the resistance of each branch of the circuit.

Ohm's Law always applies to all parts of the combination circuit

- Ohm's Law defines the relationship between voltage, current and resistance.
- $V = I \times R$ where V = voltage (EMF)
- $I = V/R$ where I = intensity of current (amperage)
- $R = V/I$ where R = resistance (load)

Summary

Combination circuits include both series and parallel circuits. Combination circuits are commonly used in electric devices and systems. Combination circuits can be analyzed by breaking the circuits down into the individual and series circuits of the combination circuit. Calculations are done on each individual circuit, then added together to get the total value for the circuit.

Concept Reinforcement

1. What is a combination circuit?
2. True or False. Ohm's Law does not apply to combination circuits.
3. List the rules for series circuits.
4. List the rules for parallel circuits.

Section 2.6 – Basic Circuit Concepts – Kirchhoff's Current Law

Section Objective

- Explain Kirchhoff's current law

A Quick Review

An electrical circuit consists of a conductor, which provides a path for electrons to flow; an energy source, which pushes the flow of electrons; and resistance, which uses the energy in the circuit to perform work.

Types of circuits:

- **Series circuit** – single path for current to flow

- **Parallel circuit** – multiple paths for current to flow

- **Combination circuit** – contains series circuit and parallel circuit

In a series circuit, current is the same throughout the circuit. Total resistance of the circuit is the sum of all of the individual resistors. The amount of voltage drop across each of the resistors adds up to equal the total voltage applied to the circuit.

In a parallel circuit, the voltage across any part of the circuit is the same as the voltage applied to the circuit. The current in each path may be different, but the total current is the sum of the current that flows in all of the branches. If you add the reciprocals of the resistance of each branch of the circuit, then the total resistance in the parallel circuit is the reciprocal of the sum of the reciprocals.

A combination circuit contains both series circuits and parallel circuits. The different types of circuits meet at junction points called nodes.

Ohm's Law defines the relationship between voltage "V," current "I" and resistance "R" in all series, parallel, and combination circuits. The relationship of a current passing through most materials is directly proportional to the potential difference applied across the material.

Law of Conservation of Charge

When current passes through a node in a combination circuit, the amount of current that flows into the node is equal to the amount of current that flows out of the node. Charge can never be created nor destroyed within an isolated system. This is called the Law of Conservation of Charge

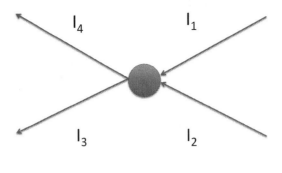

$$I_1 + I_2 = I_3 + I_4$$

Illustration of current flow through a node in a combination circuit (direct current)

Kirchhoff's Laws

In 1847 Gustav R. Kirchhoff, a German Scientist, based his work on Georg Ohm's findings. Kirchhoff created two rules for dealing with voltage and current relationships in electric circuits.

1. The algebraic sum of the currents entering and leaving a point must equal zero.
2. The algebraic sum of the voltage sources and voltage drops in a closed circuit must equal zero.

The first rule is called Kirchhoff's Current Law. The second rule is Kirchhoff's Voltage Law. We will discuss the Kirchhoff's Current Law in this section.

Kirchhoff's Current Law

Our discussions in this section are based on direct current only. Kirchhoff's Current Law is sometimes abbreviated to KCL.

When current flows through a node in an electric circuit, the amount of current flowing into the node is the same amount as that leaving the node. Current is considered to be positive as it enters the node and negative as it exits the node. This assignment of negative versus positive is purely arbitrary. The important thing is to be consistent. In a survey of various sources, the majority showed positive flow into the node and negative flow out of the node. This helps the reader keep track of the direction the current is flowing.

The Law of Conservation of Charge states that no charge is lost in an isolated (closed) circuit.

$$I_1 + I_2 = I_{total}$$

> Georg Ohm published his theory known as Ohm's Law in 1827. His work was finally recognized by the Royal Society when they awarded him the Copley Medal in 1841. According to some historians, Henry Cavendish had anticipated Ohm's Law but did not publish his findings.

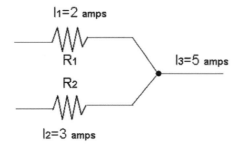

Illustration showing values of current into and out of node

Kirchhoff's Current Law states that the sum of all current in a closed circuit is zero.

$I_1 + I_2 - I_{total} = 0$

$I = 0$

Why is Kirchhoff's Current Law needed when Ohm's Law already provides the ability to analyze circuits? As circuits become more complex and contain multiple power sources, Ohm's Law becomes difficult to use in the calculation of some of the values. Kirchhoff's Current Law allows a simpler approach to solving more complex circuits.

Summary

Kirchhoff's Current Law states that sum of all current in a closed circuit is zero. In other words, the current going into the node of a combination circuit is equal to the current leaving the node. Kirchhoff's Law simplifies the process of solving complex circuits.

Concept Reinforcement

1. What is a combination circuit?

2. True or False: Georg Ohm based his work on Kirchhoff's Current Law.

3. What is Kirchhoff's Current Law?

Section 2.7 – Basic Circuit Concepts – Kirchhoff's Voltage Law

Section Objective

- Explain Kirchhoff's Voltage Law

A Quick Review

An electrical circuit provides a path for electrons to flow. A circuit includes an energy source, a conductor, and resistors.

Types of circuits:

- **Series circuit** – single path for current to flow
- **Parallel circuit** – multiple paths for current to flow
- **Combination circuit** – contains series circuit and parallel circuit

In a series circuit, the total voltage is the sum of voltage drops across each of the resistors. The sum of all of the individual resistors is equal to the total resistance. Current is the same throughout the circuit.

In a parallel circuit, the voltage applied to the circuit is the same as the voltage across any part of the circuit. If you add the reciprocals of the resistance of each branch of the circuit, then the total resistance in the parallel circuit is the reciprocal of the sum of the reciprocals. The current in each path may be different, but the total current is the sum of the current that flows in all of the branches.

When a circuit contains both series circuits and parallel circuits, it is a combination circuit. A node is the junction point where the different types of circuits meet.

Ohm's Law states that the current through a conductor between two points is directly proportional to the voltage across the two points, and inversely proportional to the resistance between them.

Law of Conservation of Energy

The voltage applied to a closed circuit is equal to the sum of the voltage drops across all the resistors. Energy can never be created nor destroyed within an isolated system. This is called the Law of Conservation of Energy.

A voltage drop is the reduction in voltage in an electrical circuit between the power source and the resistor.

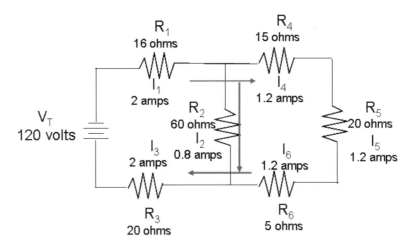

Illustration of current splitting to separate branches flowing through resistors

Kirchhoff's Laws

Based on Georg Ohm's finding, Gustav R. Kirchhoff, a German Scientist created two rules for dealing with voltage and current relationships in electric circuits.

1. The algebraic sum of the currents entering and leaving a point must equal zero.

2. The algebraic sum of the voltage sources and voltage drops in a closed circuit must equal zero.

The first rule is Kirchhoff's Current Law. The second rule is Kirchhoff's Voltage Law. We will discuss Kirchhoff's Voltage Law in this section.

Kirchhoff's Voltage Law

The discussion in this section is based on direct current only. Kirchhoff's Voltage Law is sometimes abbreviated to KVL.

Kirchhoff's Voltage Law states that the sum of the voltages around any closed circuit must equal zero. Again assume the flow of current goes from positive to negative.

In accordance with the Law of Conservation of energy, the voltage drop across each resistor (the energy used by each resistor) add up to the voltage (energy) supplied by the source.

$V_1 + V_2 + V_3 = V_{total}$

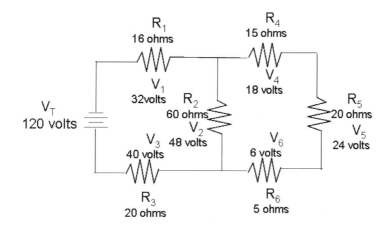

Illustration showing values of voltage drops and total voltage in circuit

Kirchhoff's Voltage Law states that the total voltage around a closed loop must be zero.

$V_1 + V_2 + V_3 - V_{total} = 0$

$\sum V = 0$

Both Kirchhoff's Current Law and Kirchhoff's Voltage Law provide a simpler approach to solving more complex electric circuits than the use of just Ohm's Law.

Summary

Kirchhoff's Voltage Law states that the sum of the voltages around any closed circuit must equal zero. The law of conservation of energy says that energy cannot be created or destroyed in a closed system. The voltage applied to a closed system is the total of the voltage drops on each of the resistors. This means that the voltage supplied to the system is equal to the voltage used by the system. A voltage drop is the result of the resistor using energy. Kirchhoff's Voltage Law provides a simpler approach than Ohm's Law to solving complex electric circuits.

Concept Reinforcement

1. What is a combination circuit?

2. What is the Law of Conservation of Energy?

3. What is Kirchhoff's Voltage Law?

Section 2.8 – Basic Circuit Concepts – Solving Circuits

Section Objectives

- Apply electrical engineering principles to solve circuits

A Quick Review

Current flows through an electrical circuit. The requirements for a closed circuit are an energy source, a conductor that provides a path for current to flow and a device that consumes energy and provides resistance to the flow of electrons.

There are three basic types of circuits:

- Series circuit which provides a single path for current to flow

- Parallel circuit which provides multiple paths for current to flow

- Combination circuit which is made up of both series circuit and parallel circuit

In a series circuit, the current is the same throughout the circuit, the total resistance is the sum of all the individual resistors and the total voltage is the sum of voltage drops across each of the resistors.

In a parallel circuit, the current flows through multiple paths. While the current may differ depending on which path it follows, the total current is the sum of the current flowing through each of the branches. The voltage applied to the circuit is the same as the voltage across any part of the circuit. If you add the reciprocals of the resistance of each branch of the circuit, then the total resistance in the parallel circuit is the reciprocal of the sum of the reciprocals.

A combination circuit contains both series and parallel circuits. A node is the junction point where the different types of circuits meet inside the combination circuit.

Ohm's Law $V = I \times R$ where V = voltage (EMF)

$I = V/R$ where I = intensity of current (amperage)

$R = V/I$ where R = resistance (load)

Kirchhoff's Current Law states that the sum of all current in a closed circuit is zero.

$I_{total} = I_1 + I_2 + I_3$

$\sum I = 0$

Kirchhoff's Voltage Law states that the total voltage around a closed circuit must be zero.

$V_{total} = V_1 + V_2 + V_3$

$\sum V = 0$

Solving Problems in Direct Current Series Circuits

In the series circuit below, the voltage provided by the battery is known to be 24 volts. The two resistors have a value of 2 ohms and 10 ohms respectively.

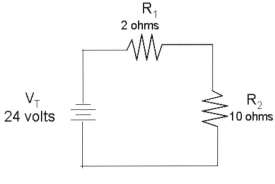

Simple series circuit

Using Ohm's Law to calculate the unknown current:

- $R_{total} = R_1 + R_2$
- $R_{total} = 2$ ohms + 10 ohms
- $R_{total} = 12$ ohms

Since V = 24 volts

- I = V/R volts/ohms
- I = 24/12 volts/ohms
- I = 2 volts/ohms since volt/ohms = amperes, then
- I = 2 amperes

If any two values are known, the third can be calculated. Because this is a series circuit, the current is the same throughout the circuit.

Solving Problems in Direct Current Parallel Circuits

In the parallel circuit below, the voltage provided by the battery is known to be 24 volts. The three resistors have a value of 15 ohms and 10 ohms and 30 ohms respectively.

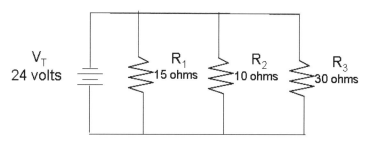

Parallel Circuit

Use Ohm's Law to calculate the unknown current.

$$R_{total} = \frac{1}{1/R_1 + 1/R_2 + 1/R_3}$$ where R_{total} is the total resistance for the parallel circuit

$$R_{total} = \frac{1}{1/15 + 1/10 + 1/30}$$ substituting values for the resistors

$$R_{total} = \frac{1}{2/30 + 3/30 + 1/30}$$ finding the common denominator

$$R_{total} = \frac{1}{6/30} = \frac{1}{0.2} = 5 \text{ ohms}$$

Since the voltage is known and the total resistance is known, the total current can be calculated:

- $I = V/R$
- $I = 24 \text{ volts}/5 \text{ ohms}$
- $I = 4.8 \text{ amperes}$

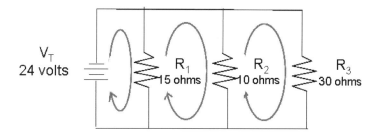

Parallel Circuit showing current flow for 3 loops

With the total voltage and resistance known, the current can be calculated for each loop inside the parallel circuit.

The first loop with a 15 ohm resistor and a 24 volt battery

- $I_1 = V/R_1$
- $I_1 = 24$ volts/15 ohms
- $I_1 = 1.6$ amperes

The second loop with a 10 ohm resistor and a 24 volt battery

- $I_2 = V/R_2$
- $I_2 = 24$ volts/10 ohms
- $I_2 = 2.4$ amperes

The third loop with a 30 ohm resistor and a 24 volt battery

- $I_3 = V/R_3$
- $I_3 = 24$ volts/30 ohms
- $I_3 = 0.8$ amperes

If the individual currents through the 3 loops are added, they should equal the total current that was previously calculated.

- $I_{total} = I_1 + I_2 + I_3$
- $I_{total} = 1.6 + 2.4 + 0.8$
- $I_{total} = 4.8$ amperes

The value of the current calculated based on the total resistance and voltage is equal to the sum of the individual currents calculated on the individual loops.

Solving Problems in Direct Current Combination Circuits

In combination circuits, the key to solving unknown values is to break the circuit into its components. In the diagram below, there is a parallel circuit inside a series circuit.

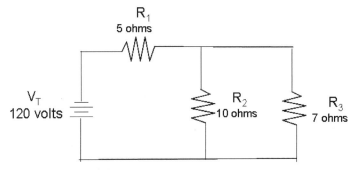

Two resistors in series and in parallel with another resistor

The voltage source is 120 volts. The two resistors in series are 5 ohms and 10 ohms. The third resistor in parallel with the two resistors is 7 ohms.

The total current through the combination circuit is found by applying Ohm's Law. But first, the total resistance must be calculated. The circuit splits with one path going through the two resistors in series and the other path going through the single resistor in parallel.

The key to solving this problem is to simplify the series portion of the circuit. The circuit follows the rules of a series circuit from the node into the series portion to the node out of the series portion of the circuit. In a series circuit, the total resistance is the sum of the individual resistors.

$$R_{series} = R_1 + R_2$$

$$R_{series} = 5 \text{ ohms} + 10 \text{ ohms}$$

$$R_{series} = 15 \text{ ohms}$$

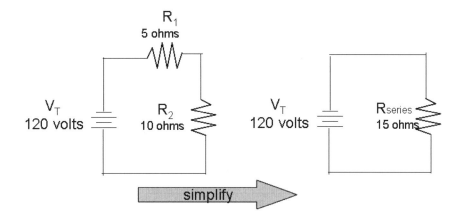

Two resistors in series converted to single resistor

Now the circuit diagram can be reduced to a simple parallel circuit.

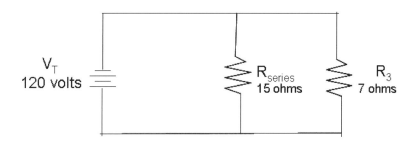

Two resistors in parallel

The parallel circuit can now be solved using Ohm's Law for the unknown values.

Summary

It is possible to solve series, parallel and combination circuits using Ohm's Law. Ohm's Law allows you to solve each circuit if you have values for two of the three variables: resistance, voltage and current. Combination circuits are solved by breaking them down into the individual series and parallel circuits, then adding the solutions for the individual circuits to get the solution for the combination circuit.

Concept Reinforcement

1. What are the three basic types of electronic circuits discussed in this section?

2. In a series circuit, if voltage is known and the value of resistors are known, what law would you use to calculate current?

3. In a combination circuit, what happens to the flow of current when it reaches a node in a parallel segment of the circuit?

Section 2.9 – Basic Circuit Concepts – Thévenin's Theorem

Section Objective

- Describe Thévenin's Theorem and its purpose in electrical circuit analysis.

A Little History

Leon Charles Thévenin (1857-1926) was a French engineer who developed the concept of equivalent electrical circuits. Leon Thévenin studied Ohm's Law and Kirchhoff's Circuit Laws. He developed a theorem whose purpose was to reduce complex circuits into simpler circuits. The theorem was named Thévenin's Theorem in his honor.

Léon Charles Thévenin

Thévenin's Theorem

Leon Thévenin reduced a circuit network into an equivalent circuit containing only a single ideal voltage source (V_0) and a single series resistor (R_0). Think of it as putting complex circuits inside an imaginary black box with two output terminals. The simple circuit is called the Thevenin Equivalent Circuit (TEC). Basically, any two terminals of a network can be reduced to one voltage source in a series with one resistor. It does not matter how many resistors, current sources and/or voltage sources are in the "black box" represented by the Thevenin equivalent circuit.

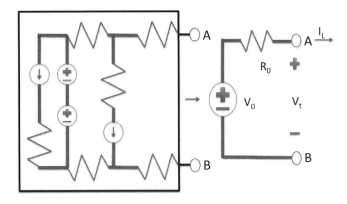

A black box and Thévenin's equivalent circuit

Thévenin's theorem is represented by the equation: $V_t = V_0 - R_0 I_L$, where V_t is the terminal voltage, V_0 is an ideal voltage source, R_0 is a resistor, and I_L is load current.

V_0 is a measure of voltage when no load is attached to the source. This is also called open circuit voltage. You can measure this by attaching a voltmeter, and nothing else, to the output terminals of the source (a battery, for example).

How do We Simplify a Complex Circuit?

There is a standard procedure for finding the Thevenin equivalent circuit.

Locate the two output terminals of the circuit

Remove any load from the terminals that is not part of the source.

Find the open circuit voltage across the terminals. This will give you the Thevenin equivalent voltage (V_{Th}).

Turn off all power sources and replace each with its characteristic resistance. Ideal voltage sources R=0, a short circuit. Ideal current sources R=∞, an open circuit.

Find the resistance across the terminals. This will give you the source resistance, R_S.

Other Considerations When Simplifying a Complex Circuit

You will often find voltage dividers when simplifying a complex circuit. A voltage divider is more complex than a TEC and can be represented by a TEC. A voltage divider has open circuit voltage, internal resistance, and a short circuit current.

Open Circuit Voltage (V_0) is the voltage of the power source before any load is attached. This is calculated using the voltage divider formula.

$$V_t = \frac{V_s \div R_a}{(R_a + R_b)}$$

Terminal voltage (V_t) is calculated by dividing the product of the system voltage (V_s) and the resistance (R_a) associated with the terminal voltage (V_0) by the sum of the resistors (R_a and R_b).

The **Short Circuit Current** is the current that would flow if a short circuit occurs. It is important to note that there is a limit to how much current a single source can supply to a load. If there is not enough current, the terminals may short circuit. The equation that represents a short circuit is: $V_0 \div R_0 = I_0$, where I_0 is the short circuit current.

Other variations on the calculations include:

- Short Circuit Current for the TEC model: $I_0 = V_0 \div R_0$
- Short Circuit Current for the Voltage Divider: $I_0 = V_s \div R_b$

The Importance of Thévenin's Theorem

The primary purpose of Thévenin's Theorem is to simplify complex networks. The theorem allows a complicated circuit containing many energy sources and many resistors, no matter how they are interconnected, to be represented by a single voltage source and a single resistor.

You can also look at this as a macromodel used for modeling electrical sources. This theory is used to explain some of what happens when you use a non-ideal power source like a battery. An example of this theorem in action is shown when you start a car that has the lights on. The lights dim as the starter is pulling energy from the battery to start the motor.

Summary

Thévenin's Theorem is used to simplify complex networks. It is used to represent complicated circuits, which contain many energy sources and resistors, with a single voltage source and a single resistor.

Concept Reinforcement

1. What is the purpose of Thévenin's Theorem?

2. What does the black box represent?

3. When Thévenin's Theorem is applied, what does the resulting circuit contain?

Section 2.10 – Basic Circuits Concepts–Norton's Theorem

Section Objective

- Explain Norton's Theorem and its purpose in electrical circuit analysis

A Little History

Edward Lawry Norton (1890-1983) was an American electrical engineer who was employed at Bell Labs at the time he published a technical memorandum on November 11, 1926 that described what is now called Norton's Theorem. A German telecommunications engineer named Hans Ferdinand Mayer published the same result in the *same month* as Norton's technical memorandum. In Europe, the theorem is known as the Mayer-Norton equivalent.

Norton's Theorem

Edward Norton reduced a circuit network into an equivalent circuit containing only a simple current source and a single parallel resistor. Norton's theorem states that any two terminals of a network can be reduced to one current source in parallel with one resistor. As with Thevenin's Theorem, it does not matter how many resistors, current sources, and/or voltage sources are in the "black box." Norton's Theorem is an extension of Thévenin's Theorem. Thévenin's black box can easily be converted to a Norton equivalent circuit using Ohm's Law.

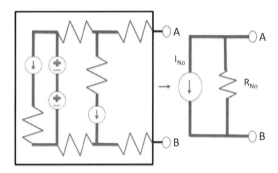

Picture of black box and Norton's equivalent circuit

Finding a Norton Equivalent Circuit.

Locate the two output terminals.

Remove anything from the terminals that is not part of the source.

Find the short circuit between the terminals. This will give you the Norton equivalent current (I_N).

Turn off all sources and replace each source with its characteristic resistance. (Ideal voltage sources R=0, a short circuit. Ideal current sources R = ∞, an open circuit).

Find the resistance across the terminals. This will give you the source resistance, R_s.

The Importance of Norton's Theorem

Like Thévenin's Theorem, Norton's Theorem is to simplify complex networks. The theorem allows a complicated circuit containing many energy sources and resistors to be represented by a simple current source and a single parallel resistor.

Thevenin's Theorem and Norton's Theorem represent two parts of a duality that is common in electrical engineering. A duality occurs when one variable usually appears with another variable. In this case, voltage and current are duals. The dual of the Thevenin Equivalent model is the Norton Equivalent Model. The result of this is that the Thevenin Equivalence Circuit (TEC) is electrically equivalent to the Norton Equivalence Circuit (NEC).

There are some important similarities between the Thevenin and Norton equivalents:

R_0, internal resistance, is the same in both types of circuits.

$V_0 = R_0 I_0$ where V_0 represents open circuit voltage in the Thevenin model, and I_0 is the short circuit current in the Norton model.

Both Theorems follow Ohm's Law, meaning that if any two quantities are known, the third can be determined.

Summary

The Norton Equivalence Theorem is used to simplify complex circuits to a black box and a Norton's Equivalent Circuit. A Norton's Equivalent Circuit contains a simple current source and a parallel resistor. This Theorem is closely related to Thevenin's Equivalence Theorem. Both simplify complex circuits to a black box and an equivalence circuit with a single source and a single resistor. The TEC has a source and resistor in series. The NEC has a source and parallel resistor.

Concept Reinforcement

1. What is the purpose of Norton's Theorem?

2. When Norton's Theorem is applied, what does the resulting circuit contain?

3. True or False. Thévenin's black box cannot be converted to a Norton equivalent circuit.

Section 2.11 – Basic Circuit Concepts – Superposition Theorem

Section Objective

- Describe the Superposition Theorem

A Little History

If you could combine both of Kirchhoff's Laws, Thevenin's Theorem and Norton's Theorem, you would get something like the Superposition Theorem. No one person is credited with the development of the Superposition Theorem.

The Superposition Theorem

Suppose you have a circuit containing multiple energy sources. Refer to the example shown below.

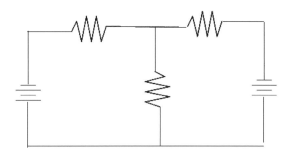

A circuit with two energy sources

The Superposition Theorem states that the current in any branch of a circuit that has multiple energy sources will be the algebraic sum of the individual currents produced by each of the energy sources. In a circuit that has two energy sources, the procedure for solving the circuit is to pretend that one of the sources doesn't exist (replace it with a wire). The circuit is then a combination circuit and the total current flow for the circuit and the individual current flows through the parallel branches can be calculated. The next step is to add the missing energy source back in and pretend the first energy source does not exist. Again, the circuit becomes a combination circuit with the second energy source. The current for the second energy source can be calculated for both the total circuit and the parallel branches. The current total will always be the same in a closed circuit. The Superposition Theorem shows how to calculate the flow through a parallel branch that has multiple energy sources in the circuit by summing the results of the independent calculations based on each energy source.

Let's break this down into a list of steps.

- Replace all but one of the voltage sources with a short (a wire, for example).
- Replace all of the current sources with an "open."
- Solve for the current going through each resistor.
- Change the voltage source.
- Repeat steps 1-3 until all the voltage sources have been used.
- Add all the currents for each resistor.
- Label the original circuit.

This sounds more complicated than it is, so let's walk through an example. The following example shows how the calculations are performed on a circuit containing two energy sources and two resistors in parallel with a third resistor.

First eliminate one of the energy sources as shown in below. Complete the calculations as if that energy source V_2 was non-existent. Make a record of those values.

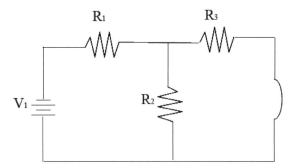

A circuit where one of the energy sources is replaced with a wire

Since this is now a combination circuit (without V_2), the resistance is calculated by figuring out the resistance of the parallel portion of the circuit and adding it to the series portion of the circuit.

$$R_{total-1} = R_1 + R_{parallel-1}$$

$$R_{parallel-1} = \frac{1}{\frac{1}{R_2} + \frac{1}{R_3}}, \text{ where } R_2 \text{ and } R_3 \text{ are in parallel}$$

$$R_{parallel-1} = \frac{1}{\frac{1}{150} + \frac{1}{600}}, \text{ substituting the known resistor values}$$

$$R_{parallel\text{-}1} = \frac{1}{\frac{4}{600} + \frac{1}{600}}, \text{ solving the equation}$$

$$R_{parallel\text{-}1} = \frac{1}{\frac{5}{600}}, \text{ solving the equation}$$

$$R_{parallel\text{-}1} = \frac{1}{0.0083}, \text{ solving the equation}$$

$$R_{parallel\text{-}1} = 120 \text{ ohms}$$

Add the parallel portion of the circuit to the series portion to calculate the total resistance of the circuit for V_1. Remember that for this calculation, V_2 does not exist.

$$R_{total\text{-}1} = R_{parallel\text{-}1} + R_1$$

$$R_{total\text{-}1} = 120 \text{ ohms} + 300 \text{ ohms}$$

$$R_{total\text{-}1} = 420 \text{ ohms}$$

Since the current would be the same throughout the circuit, the total current can be calculated:

$$I_{total\text{-}1} = V_1/R_{total}$$

$$I_{total\text{-}1} = 12 \text{ volts}/420 \text{ ohms}$$

$$I_{total\text{-}1} = 0.02858 \text{ amps}$$

The next step is to calculate the voltage drop across the parallel portion of the circuit as shown in the image below.

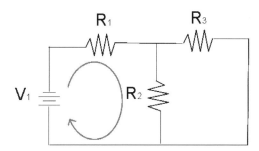

The circuit is shown as a combination circuit without V_2

$$V_{parallel-1} = I_{total-1} \times R_{parallel-1}$$

$$V_{parallel-1} = 0.0286 \text{ amps} \times 120 \text{ ohms}$$

$$V_{parallel-1} = 3.432 \text{ volts}$$

Since the total current is equal to 0.02858 amps, the current through R_2 (part of the parallel circuit) can now be calculated:

$$I_{2 \text{ for } V1} = \frac{V_{parallel-1}}{R_2}$$

$$I_{2 \text{ for } V1} = \frac{3.432 \text{ V}}{150 \Omega}$$

$$I_{2 \text{ for } V1} = 0.0229 \text{ amps}$$

Note the flow of current through the R_2 is shown by the arrow in image above.

The direction of the flow of current is important when we total the flow through the resistor that is contributed by each of the energy sources. The current for each energy source will be summed if current flows through R_2 in the same direction.

Now reinstate the second energy source and eliminate the first energy source. Perform the calculations and make a record of the second set of values.

$$R_{total-2} = R_3 + R_{parallel-2}$$

$$R_{parallel-2} = \frac{1}{\frac{1}{R_1} + \frac{1}{R_2}} \text{, where } R_1 \text{ and } R_2 \text{ are in parallel}$$

$$R_{parallel-2} = \frac{1}{\frac{1}{300} + \frac{1}{150}} \text{, substituting the known resistor values}$$

$$R_{parallel-2} = \frac{1}{\frac{1}{300} + \frac{2}{300}} \text{, solving the equation}$$

$$R_{parallel-2} = \frac{\frac{1}{3}}{300} \text{, solving the equation}$$

$$R_{parallel-2} = 100 \text{ ohms}$$

$R_{total-2}$ = 600 ohms + 100 ohms

$R_{total-2}$ = 700 ohms

The total current for the circuit with V_2 energy source only can be calculated:

$I_{total-2} = V_2/R_{total-2}$

$I_{total-2}$ = 6 volts/700 ohms

$I_{total-2}$ = 0.00857 amps

The next step would be to calculate the voltage drop across the parallel portion of the circuit as shown below.

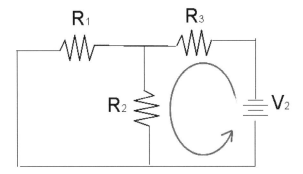

The circuit is shown as a combination circuit with V_2.

$V_{parallel-2} = I_{total-2} \times R_{parallel-2}$

$V_{parallel-2}$ = 0.00857 amps × 100 ohms

$V_{parallel-2}$ = 0.857 volts

Since the total current is equal to 0.00857 amps, the current through R_2 (part of the parallel circuit) can now be calculated:

$I_{2 \text{ for V2}} = \dfrac{V_{parallel-2}}{R_2}$

$I_{2 \text{ for V2}} = \dfrac{0.857 \text{ v}}{150 \Omega}$

$I_{2 \text{ for V2}}$ = 0.00571 amps

Note the flow of current through the R_2 is shown by the arrow in image above.

The direction of the flow of current is important when we total the flow through the resistor that is contributed by each of the energy sources. The current for each energy source will be summed if current flows through R_2 in the same direction. Since it does in this case, the total current through R_2 is the algebraic sum of the individual currents calculated for each energy source.

$I_{2total} = I_{2forV1} + I_{2forV2}$

$I_{2total} = 0.0229$ amps $+ 0.00571$ amps

$I_{2total} = 0.0286$ amps

Remember that the total current flow through out the circuit is the same at any node. This method provides the tools to calculate the current through a branch (in this example R_2) with multiple energy sources.

While this may be a lengthy process, it is a simple way to analyze complex linear circuits with multiple energy sources. The Superposition Theorem for electrical circuits states that the total current in any branch of a linear circuit equals the algebraic sum of the currents produced by each source as if it were acting alone throughout that particular circuit.

The Importance of the Superposition Theorem

The Superposition Theorem is used to convert any circuit into its Thevenin equivalent or its Norton equivalent. Thevenin's Theorem and Norton's Theorem both convert complex circuitry into a virtual black box. Thevenin's black box is a single voltage (energy) source and a single series resistor with two terminals. Norton's black box is simple current source and a single parallel resistor with two terminals.

Summary

In summary, to calculate the contribution of each individual source, all of the other sources first must be eliminated by setting them to zero.

This procedure is followed for each source in turn. The resultant currents are added to determine the true operation of the circuit. The resultant circuit operation is the superposition of the various energy sources.

Because power dissipations are nonlinear functions, power cannot be determined by adding the power related to the individuals sources. Also, if a resistor changes its value or the amount of current that flows through it over time, it is nonlinear. In these cases the Superposition Theorem cannot be used.

Concept Reinforcement

1. Explain the process of calculating total current using the Superposition Theorem.

2. True or False. The Superposition Theorem is used to convert any circuit into its Thevenin equivalent or Norton Equivalent.

3. Explain how the Superposition Theorem relates to Kirchhoff's Laws, Norton's Theorem, and Thevenin's Theorem.

Section 2.12 – Basic Circuit Concepts – Millman's Theorem

Section Objective

- Explain Millman's Theorem

A Little History

Jacob Millman was born in Russia in 1911. He received his PhD from MIT in 1935 and was a professor of electrical engineering at Columbia University. Millman's Theorem is named after him. Millman's theorem has also been called the parallel generator theorem and is a means to simplify the solution of a circuit.

A Quick Review

In order to understand Millman's Theorem, we need to review the basics of circuitry.

A simple electric circuit is the flow of electrons (I) through a conductor which contains an energy source (V), some form of resistance (R) and a return to the energy source. When the electrons flow in one direction, the circuit is direct current (DC).

Ohm's Law defines the relationship between energy (V), resistance (R) and current (I) and provides a means to calculate current when energy and resistance is known. Alternately, it provides the means to calculate energy when resistance and current are known and resistance when energy and current are known.

$I = V/R \qquad V = I \times R \qquad$ and $R = V/I$

Additional energy sources, resistors, and other devices can be included in the circuit. This just makes the circuit more complicated and the calculation of values requires special tools. These special tools come in the form of rules and theorems which provide a method of simplifying the circuit and enabling the calculation of unknown values.

The following is a review of some basic rules for analyzing circuits.

In a series circuit there is only one path for the current to flow therefore the current is the same throughout the circuit. The total resistance in a series circuit is found by simple addition of the individual resistors. Because of the Law of Conservation of Energy, voltage drops across each resistor in the circuit must add up to equal the amount of energy applied to the circuit.

In a parallel circuit, the current splits between the branches. The current does not split evenly unless the resistance in each branch is equal. Since there are multiple paths for the currents to flow, the measurement of the current may not be the same at any point in the circuit; however, the total amount of current is equal to the sum of the currents flowing through each parallel branch. The voltage across any part of the circuit is the same as the energy applied to the circuit. The calculation of the total resistance in a parallel circuit is the reciprocal of the sum of the reciprocals of the individual branches. While the calculation of the total resistance is a little more complicated than in a series circuit, it is not difficult, just more time consuming and care has to be taken to not skip steps in the calculations in order to avoid mistakes.

A combination circuit contains both parallel and series branches. The junction point of the branches is called a node. To calculate values in a combination circuit, the circuit is broken up into individual segments and the rules for calculations apply to the segment.

To make complex circuits easier to calculate, Kirchhoff's Laws provides two rules for dealing with voltage and current relationships. Kirchhoff's Current Law states that the algebraic sum of the currents entering a node and leaving a node must equal zero. This is based on the Law of Conservation of Charge. Kirchhoff's Voltage Law states that the algebraic sum of voltage (energy) sources and voltage drops in a close circuit must equal zero. This is based on the Law of Conservation of Energy.

Because circuits can get very complicated and the ability to analyze them becomes almost impossible, some very clever scientists and engineers have figured out additional ways to simplify the circuits so that calculations can be performed.

One of these is Thevenin's Theorem. In this theorem, an equivalent circuit is developed that contains only a single voltage (energy) source and a single series resistor. All of the complex circuitry is hidden inside a virtual black box with two terminals.

Another of these simplification theorems is Norton's Theorem. Norton developed an equivalent circuit containing only a simple current source and a single parallel resistor. Again the complex circuitry is hidden inside a virtual black box with two terminals.

Using Ohm's Law, Thevenin's black box could be converted to a Norton black box.

By understanding the concept of putting complex circuits inside a virtual black box and dealing only with the input and output of the black box, the black box becomes simply a building block in circuits.

Using the black box concepts developed by Thevenin and Norton, Jacob Millman developed an equation to treat each branch in a circuit (with its series voltage source and resistance) as a Thevenin equivalent circuit. Then each Thevenin equivalent circuit is converted to an equivalent Norton circuit (Ohm's Law applies). This conversion matched with a parallel resistance formula enables the calculation of the total voltage across all branches of the circuit.

Millman's Theorem

Millman's Theorem only applies to those circuits that can be drawn as a set of parallel connected branches as shown in the following image.

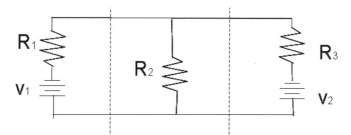

Apply Millman's Theorem by dividing the circuit into three segments.

Millman's Theorem is an equation that will provide the means to calculate the voltage across all of the branches.

$$V_{across\ all\ branches} = \frac{\frac{V_1}{R_1} + \frac{V_2}{R_2} + \frac{V_3}{R_3}}{\frac{1}{R_1} + \frac{1}{R_2} + \frac{1}{R_3}}$$

Summary

In summary, Millman's Theorem provides a simple method to compute the voltage at the ends of a circuit that consist of only branches in parallel. It is based upon several other basic theorems of electrical engineering, including Ohm's law, Kirchhoff's Theorems, and Thevenin's and Norton's theorems of equivalent circuits.

Concept Reinforcement

1. True or False. Millman's Theorem provides a means to compute the voltage in series circuits.

2. True or False. Millman's Theorem uses Kirchhoff's Current Law to reduce complex circuitry to a virtual black box.

3. Write Millman's Theorem equation for calculating voltages across all of the branches of a 3 branch circuit. Solve for total voltage across all the branches.

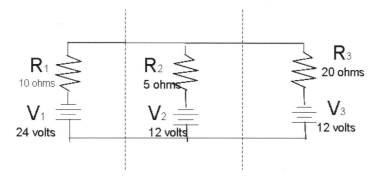

Circuit with three parallel branches containing resistors and energy sources

119

Section 2.13 – Measuring Instruments

Section Objective

- Describe instruments used to measure electricity

Background

Calculating electric circuit values using Ohm's Law, Kirchhoff's Current and Voltage Laws, Thevenin's Theorem, Norton's Theorem, the Superposition Theorem and any other laws or theorems out there is very valuable in understand what is happening in a circuit. It is certainly the basis for design and planning to build circuits. However, it is important once electric circuits are a reality to be able to actually measure what is happening inside the circuit.

You may ask why. After all, the circuit is built. Why is it important to measure electrical values? What's the point? Being able to measure the actual operation of a circuit provides the means by which to analyze what is actually happening. If the device is not performing in the expected manner, then measuring the circuit at key points and comparing the values to those expected using the design schematics, provides the engineer or technician the means to determine if there is a component failure or if there is a design flaw.

Digital Meters and Analog Meters

A clock that has hands that go around in a circle is an analog instrument that measures time. An alarm clock that displays digital numbers is a digital alarm clock that measures time (as well as waking you up). Analog Meters use a pointer and a scale to provide the measurement being tested. Like a digital alarm clock, digital meters use actual numbers on a display to provide the reading of the measurement.

The Voltmeter

The voltmeter measures energy and is connected directly across the energy source. The voltmeter is designed to test the voltage of an energy source such as a battery.

Voltmeter attached to an energy source

Voltmeters can have a digital readout as shown in below.

Digital Voltmeter

The dials and switches on the voltmeters provide the ability to read a wide range of values. These voltmeters are called multi-range voltmeters. By switching the scale using the dials or switches on the front of the voltmeters, the range of the meter can be selected. This widens the number of applications that a single voltmeter can analyze.

Voltmeters can have analog readouts as shown below.

Analog Voltmeter – Demonstration voltmeter from a physics class–Wikipedia

A voltmeter can be put in series with a resistor. Before the resistor value can be computed, the meter's operating characteristics must be taken into account since the voltmeter changes the resistance of the circuit. This information should be provided in the operations manual for the voltmeter.

The Ammeter

The ammeter is used to measure current in an electric circuit. The ammeter must be connected in series with the load. Like voltmeters, ammeters can be multi-range.

Some circuits may require shutting down before the ammeter can be installed since the ammeter needs to be in series to read the current. There is one type of ammeter that most electricians use because it does not have to be wired into the circuit. It is a clamp on meter.

Clamp on ammeter.

The clamp-on meter clamps directly around one of the conductors supplying power (energy) to the load (resistance). If the meter is clamped around both the supply and the return line, the magnetic fields of the wires cancel each other and the reading is zero. This does not mean there is no current in the wire. This means it is not connected properly. The clamp-on meter clamps directly around one conductor only. You need to be careful when clamping a device around a wire and do not clamp the wire itself. That could damage the wire. If you managed to penetrate the shielding, you could provide a path for those nasty little electrons to get out and cause you a bit of a shock. The watchword with working around anything with electric current flowing through it is "BE CAREFUL." If you don't know what you are doing, don't do it. Ask for help.

The Ohmmeter

The ohmmeter is used to measure resistance in an electric circuit. The ohmmeter also comes in analog or digital. However, they function differently. Analog meters operate by measuring the amount of current change in the circuit when an unknown value of resistance is added.

Analog ohmmeter–Wikipedia

Digital ohmmeters measure resistance by measuring the amount of voltage drop across an unknown resistance.

Like voltmeters and ammeters, ohmmeters can be multi-range.

Meter Reading

The steps to using and reading any meter are:

- Select the proper meter for the job (what are you trying to measure?)
- Read the directions on how to use the meter (unless you have already been thoroughly trained in its use)
- Connect the meter correctly to the circuit to read the values of interest (follow the instructions in the operations manual)
- Make sure the range is properly set for the reading
- Read the meter
- Record the results

Don't forget to record the results. After all, the reading is the reason for using the meter. However, each step in the process is equally important.

Summary

Meters are used to measure electricity. They measure the electrical activity within a circuit. There are several types of meters. Analog meters use a pointer and a scale to measure electricity. Digital meters use the actual numbers on a display to show the measurement. Voltmeters measure voltage. The Ammeter measures amps and the Ohmmeter measure ohms. It is important to select the correct meter for the measurement you wish to take.

Concept Reinforcement

1. What type of meter is used to measure resistance?
2. What type of meter is used to measure current?
3. What type of meter is used to measure energy?
4. What is the main difference between digital and analog meters?

Section 2.14 – The Wheatstone Bridge

Section Objective

- Explain a Wheatstone Bridge and its uses

A Little History

Samuel Hunter Christie invented a method to measure electrical resistance. He published his diamond method in 1833. Ten years later, Charles Wheatstone presented an improved version of Christie's work. Even though Charles Wheatstone gave full credit to Samuel Hunter Christie, the concept was associated with Wheatstone because of his practical applications.

Sir Charles Wheatstone

The Wheatstone Bridge

The purpose of the Wheatstone bridge is to measure resistance. It provides a precise comparison of resistances. It is very useful for measuring small changes in resistance, which makes it suitable for measuring the resistance in a strain gauge. The Wheatstone bridge has a source of energy and a galvanometer that connects two parallel branches. A galvanometer is a type of ammeter. It indicates the direction of the flow of current as well as the amperage. The two parallel branches contain three known resistors and one unknown resistor. The three known resistors are adjusted and balanced until the current passing through the galvanometer goes to zero.

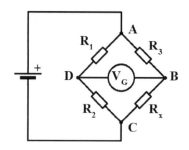

Wheatstone bridge circuit–Wikipedia

A sample of a Wheatstone bridge is shown below.

Wheatstone bridge

The Wheatstone bridge operates on the principle that the applied voltage is equal to the sum of the voltage drops in a series circuit. It is most commonly used to determine an unknown resistance.

The bridge is balanced when there is no voltage difference between the terminals. In this condition, the resistance ratio between the series resistors in each branch are equal. If there is a difference of potential between the two branches, voltage develops.

Summary

The Wheatstone Bridge is a tool used to measure resistance in electrical circuits. The branches of the circuit are analyzed for voltage potential, which may be different between the terminals. If the terminals are balanced, the voltage potential is zero.

Concept Reinforcement

1. True or False. A galvanometer measures resistance.

2. True or False. A Wheatstone bridge is used to make accurate measurements of resistance and operates on the principle that the sum of the voltage drops in a series circuit must equal the applied voltage.

3. How many known resistors are there in a Wheatstone bridge?

4. How many unknown resistors are there in a Wheatstone bridge?

Section 2.15 – An In-Depth Review of Circuit Concepts

Section Objectives

- Explain the different types of electrical circuits and how to solve circuit problems

- Describe electrical schematics, electronic symbols and the terminology used in circuit design

- Discuss circuit solving theorems and instruments used to measure electricity

A Quick Review

Electrical schematics are drawn using electrical symbols. From wiring layouts for construction builders to circuit board layouts for computers, schematics are the building plans for electrical work. Schematics are also used to document work that has been completed and can be used to debug or solve electrical problems.

Automobiles now have very complex electrical systems. The trained automobile mechanic must now perform the work of an electrical technician. If something goes wrong with the electrical system in a car, the technician must use diagnostic tools to determine the problem. Diagnostic tools can include schematics, measuring instruments and in many cases a complex computer program that runs on a specialized computer system designed for that particular automobile.

A number of laws and methods are available for understanding and analyzing electric circuits:

- Law of Charges – likes attract, opposites repel

- Law of Centrifugal Force- rotating object pulls away from its center and increases in force the faster it spins

- Law of Conservation of Charge – the amount of current can never be created nor destroyed in an isolated system

- Law of Conservation of Energy – energy can never be created nor destroyed within an isolated system

- Ohm's Law – the amount of current is directly proportional to the amount of energy applied and inversely proportional to the amount of resistance in a closed circuit

- Kirchhoff's Current Law – the algebraic sum of currents entering and leaving a node in a closed circuit equal zero

- Kirchhoff's Voltage Law- the algebraic sum of the voltage sources and the voltage drops in a closed circuit equal zero

- Thevenin's Theorem – reduces a complex circuit into an equivalent virtual black box containing a single voltage source and a single series resistor with two output terminals

- Norton's Theorem – reduces a complex circuit into an equivalent virtual black box containing a single current source and a single parallel resistor with two output terminals

- Superposition Theorem – the current in any branch of a circuit with multiple energy sources, is the algebraic sum of the individual currents produced by each energy source.

- Millman's Theorem Equation – provides a method to calculate the voltage across all of the branches in a circuit containing only parallel branches

The following are types of circuits in a closed loop direct current environment:

- Series Circuits
  - Current flows in one path
  - The current is the same at any point in the circuit
  - The total resistance is the sum of the individual resistors
  - The amount of voltage drops across all the resistors must equal the voltage applied to the circuit

- Parallel Circuits
  - There are multiple paths for current to flow
  - The total current is the sum of the current that flows through all the branches in the circuit
  - The voltage across any part of the circuit is the same as the applied voltage
  - The total resistance is the reciprocal of the sum of the reciprocals of the resistance of each branch of the circuit

- Combination Circuits
  - Consists of both series and parallel circuits
  - The circuit can be divided up into series and parallel segments
  - The rules for a series circuit apply to the segment of the circuit that is in series
  - The rules for a parallel circuit apply to the segment of the circuit that is in parallel

The following instruments are used to verify the actual values in a circuit:

- Voltmeter – measures the amount of voltage in an energy source
- Ammeter – measures the amount of current flowing through a conductor
- Ohmmeter – measure the resistance in an electric circuit

The following contains some of the electronic symbols used in electrical schematics.

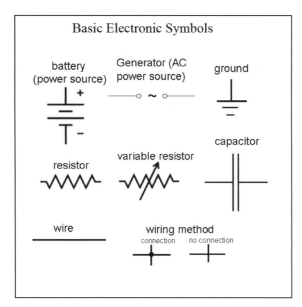

Electronic Symbols

An Exercise for You

The image below shows a schematic of a circuit containing 3 parallel paths containing resistors and power sources. Calculate the total current flowing through R_2 using the Superposition Theorem.

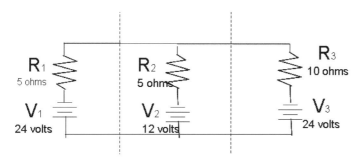

Combination circuit with multiple energy sources

Eliminate each of the energy sources one at a time and calculate the current that flows through the system for each energy source. You will find a value for the current that flows through R_2 for each calculation. Add the currents for the total current that flows through R_2 when all energy sources are in place and connected.

What is the total current flowing through R_2?

Summary

Electrical Engineering uses well-established theorems and protocols for describing and analyzing electrical circuits. Electrical circuits are included in almost everything we use, from kitchen appliances to cars. Our computers are full of circuits, which perform the calculations, store data, and perform the basic operating functions required for the computer to work. In order to be successful, an electrical engineer must be completely familiar with the symbols and protocols used to draw electrical circuits.

Concept Reinforcement

1. Name the three types of circuits discussed.

2. What is the electronic symbol for a resistor (resistance)?

3. What instrument is used to measure current?

4. What laws and theorems did you use in the problem for this lesson?

Unit Three

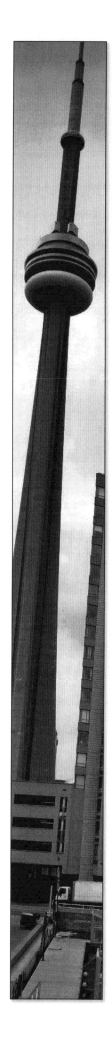

Section 3.1 – How Things Work – On/Off Switches   129

Section 3.2 – How Things Work – Lights and Lighting   137

Section 3.3 – How Things Work – Portable Lights   149

Section 3.4 – How Things Work – Car Batteries   157

Section 3.5 – How Things Work – Water Pump in an Internal Combustion Engine of an Automobile   167

Section 3.6 – How Things Work – Ceiling Fans   173

Section 3.7 – How Things Work – Portable Electric Heaters   179

Section 3.8 – How Things Work – Microwave Ovens   189

Section 3.9 – How Things Work – Wireless Remote Controllers   195

Section 3.10 – How Things Work – Safety Sensors (Carbon Monoxide Sensors)   203

Section 3.11 – How Things Work – Direct Current Motors   211

Section 3.12 – How Things Work – Alternating Current Motors   225

Section 3.13 – U.S. and International Electrical Standards   233

Section 3.14 – How Things Work -Power Transmission and Alternative Sources   243

Section 3.15 – Applications of Electrical Engineering   253

Section 3.1 – How Things Work – On/Off Switches

Section Objective

- Explain how an on/off switch works

Background Information

An **electric circuit** contains a power source, a conductor to carry electric current and a resistance (load).

Electric current (I) is the flow of electrons along a conductor.

An **energy source (V)** provides force to move the electrons along a conductor.

A **resistance (R)** consumes energy and is the load on a system.

Direct current (DC) is current that flows only in one direction from the energy source and back. Batteries provide direct current.

Alternating current (AC) is current that switches direction periodically. The movement (or flow) of electrons periodically reverses direction. Most power supplied to homes and businesses is alternating current.

A **schematic** is an electrical blueprint (drawing) of a circuit design.

What is a Switch?

A switch is made of two pieces of conducting metal. The two pieces of medal are called contacts. When the contacts touch, the circuit is closed and current will flow. When the two pieces do not touch, the circuit is broken (open) and no current will flow. When an external force is applied to the switch, the switch will change state. Depending on the type of switch, it will either open or close the contacts.

Function of an On/Off Switch

Current is created by the flow of electrons. An energy source pushes the electrons to flow along a path provided by a conductor. When the electrons meet a resistance, some of their energy is consumed by the resistance (load). When an energy source, a conductor and a resistor are connected, it becomes a circuit. Current will flow as long as there is a continuous path along a conductor with no break or interruption in the path with a return to the energy source. The image below shows a simple closed electric circuit. There is an on/off switch in the circuit in the closed position allowing current to flow.

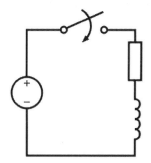

Simple electric circuit with a switch that will open and close.

Putting the switch in the open position stops the flow of electrons and there is no flow of current in the circuit. This is called an open circuit. Any electrical device in this circuit will shut off.

A switch in an electrical circuit breaks the circuit by interrupting the flow of electrons when the contacts are open. The switch permits current to flow by closing the circuit when the contacts are closed. The **primary purpose of a switch** in an electrical circuit is to either turn on or shut off power to electrical devices.

Types of Switches

Simple **on/off switch** installed in an electrical circuit will interrupt the circuit (create an open circuit) when in the off position and will provide a path for current to flow (create a closed circuit) when in the on position.

Rocker dip switch

There are many different types of switches. The following is a partial list:

- Momentary switches – push to make and push to break
- Pushbutton switches
- Rocker switches
- Toggle switches
- Dip switches
- Rotary switches
- Pull-chain switches

- Timer switches
- Voice-activated switches
- Sound-activated switches
- Slide switches
- Mercury switches
- Magnetic switches
- Knife switches
- Latching switches

All of these switches have one thing in common. They make or break current flow in a circuit.

Definitions:

CONTACTS are the two pieces of conducting metal that touch to make a circuit, and separate to break the circuit.

OPEN means the contacts are not conducting indicating an open circuit and is considered to be in the **off** position.

CLOSED means the contacts are conducting indicating a closed circuit and is considered to be in the **on** position.

POLE is the number of contact sets in a switch. A simple on/off switch will have one pole consisting of one set of contacts.

THROW has one or two conducting paths that can be made by the switch. For a single throw, there is only one path that conducts current. For a double throw, there is a choice of two paths that can be made by the closure of the contacts. In the case of the double throw, current will flow down the path that is chosen by the switch and not the other path. When the switch changes state, current will switch paths.

WAY is the number of conducting paths that can be made by the switch. One contact can switch between multiple contacts (connections). The direction of the current flow will depend upon which contact (connection) is made.

MOMENTARY switch is a switch that returns to its "normal" state when released. Some switches are normally open or **off** (i.e. no current flows) unless pushed. Doorbells are usually momentary push to close switches. When pushed, the contact is made and the door bell rings. When the door bell switch is released, the contact is open and the door bell stops ringing. Alternately, some switches are normally closed or **on** (i.e. current flows) unless the switch is pushed.

Switches can be shown on electrical schematics using symbols. A few common symbols are shown in table 1.

Table 1

Electronics abbreviation	Definition of abbreviation	Description	Symbol
ON-OFF SPST	single pole, single throw	on-off switch	
Dual ON-OFF DPST	double pole, single throw	pair of switches that operate together	
ON-ON SPDT	single pole, double throw	changeover switch	
Dual ON-ON DPDT	double pole, double throw	pair of ON-ON switches that operate together	
(ON)-OFF SPST Momentary	push-to-make momentary	returns to normally open (OFF) position when button is released	
ON-(OFF) SPST Momentary	push to break momentary	returns to normally closed (ON) position when button is released	
MULTI-WAY	single pole 4 way switch	Multi-switch (many switch positions)	

Why are Switches Important?

Without switches it would be difficult to turn electrical devices on and off. If electrical devices such as lamps, radios, TVs and computers were left on when not in use, it would be annoying and waste energy but not dangerous. However, some devices if left on indefinitely would overheat and could cause a fire. Sometimes it is necessary to shut something off quickly for safety reasons. Imagine a chain saw that does not have an off switch. Setting it down to go unplug it could be dangerous.

Applications

Single pole single throw (SPST) wall switch operated on household alternating current

Single pole single throw switches are the most common in a household. It is labeled "on" on one side and "off" on the other side.

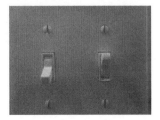

Wall switches are wired into an electrical circuit (alternating current – not direct current), which is connected to one of the circuit breakers.

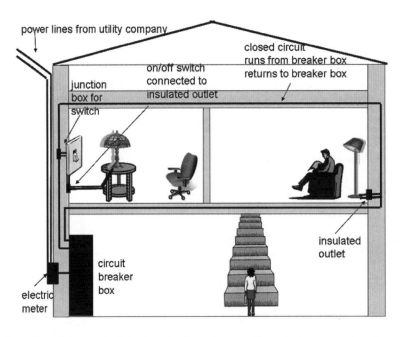

Simplified household circuit showing wall switch operating lamp

The wall switch is mounted inside a well-insulated box and all metal parts where excess current could flow are grounded. The box is designed to contain a switch that is safe for turning electrical devices on and off. Most applications of the single pole single throw wall switch are to turn a light on or off.

Insulated box containing wall switch

A typical household wall switch tied into an incandescent light, has a strip of conducting metal. When the switch is moved to the "ON" position, the metal strip (known as a contact) connects with the second contact, completing the circuit. Current flows through the circuit providing electrons which heat the filament inside the light bulb and "turn on" the light. This is the "ON" position of the switch and the circuit is "closed."

The external switch is really an insulated lever. When the switch is moved to the "OFF" position, it pushes the metal conducting strip away from the contact and breaks the circuit. The circuit is "open," the switch is "OFF," no current flows and the light "turns off." This type of switch is a single pole single throw switch and because it uses a lever to open/close the contacts, it is called a toggle switch. A wall switch is called a wall switch because it is mounted in an insulated box in the wall.

The application of the wall switch described in this lesson uses alternating current. Single pole single throw (SPST) switches are used in both direct current (battery operated) circuits and alternating current (household) circuits. The concept of making contact between two conducting metal strips to permit current to flow and breaking contact to stop current from flowing is the key to the usage of any switch.

Single pole single throw battery operated doorbell

The doorbell shown below is not tied into the household wiring. It is battery operated. One advantage of this is that the doorbell will operate during a power outage. Also, no physical wiring is required resulting in an easier and less expensive installation of the battery operated door bell. The disadvantage is that if the battery loses power and is not replaced, the doorbell will not operate.

Picture of door bell

The doorbell uses a different type of switch. While the switch is a single pole single throw switch similar in concept to the wall switch, it is a push-to-make momentary switch. The circuit will only stay closed while the doorbell switch is being pushed. When the circuit is closed, the bell which is powered by the current from the battery will sound. As soon as the doorbell is no longer being pushed, the circuit will break and the sound will stop.

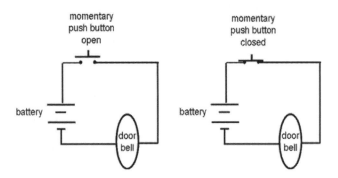

Wiring diagram showing open circuit on left (doorbell not ringing) and closed circuit on right (doorbell ringing)

Summary

The basic concept of switches is simple. The primary purpose of a switch is to provide a means to shut off power to electrical devices. By opening a circuit, a switch stops the flow of electrons. Any electrical device in that circuit is "shut off". No flow of electrons means there is no current and that means there is no power to operate electric devices.

While the concept is simple (open or close a circuit), the number of different types of switches is extensive. Selecting the right switch for the job is important. The switch must be able to do the job (carry the right amount of current when closed and stop the flow of current when open). The switch must be made of the right materials for its usage. Switches are rated for maximum voltage and current. It is not safe to exceed the maximum rated voltage and current through a switch.

The type of switch implemented also depends upon the application. An on/off switch for a doorbell would be extremely annoying since it would not be silenced until the switch was physically turned to the off position.

Concept Reinforcement

1. What is the primary purpose of a switch?

2. True or False. Switches only work in direct current circuits.

3. When a switch is "on," is the circuit open or closed?

4. What is the difference between an on/off switch and an on/off momentary switch?

Section 3.2 – How Things Work – Lights and Lighting

Section Objective

- Explain how lights and lighting fixtures work

Background Information

Electrical power is measured in watts. A more common term of power measurement is Megawatts (MW).

1 Megawatt = 1 million watts

Electrical energy is measured in watt-hours. A more common term for energy measurement is kilowatt-hour (KWH).

I kilowatt = 1000 watt-hours

A light bulb with a **power rating** of 60 watts indicates the instantaneous rate at which it converts electrical energy. If the bulb is turned on for one hour, it will consume a total of 60 watt-hours of energy into heat and light.

Valence electrons are electrons that travel in the outer orbit of an atom. When the atom is bombarded, it gives up its valence electrons. The maximum number of valence electrons in an outer orbit is 8. The fewer the valence electrons in the outer orbit, the more likely the atom is to give them up. **Electric current** in carried by the flow of electrons through a conductor from a negatively charged area to a positively charged area.

An ion is an atom that has lost or gained an electron and therefore has a positive or negative charge. Ions are drawn to oppositely charged areas. If the ion is negative (it has gained an electron), it moves towards a positive area. If the ion is positive (it has lost an electron), it moves towards a negative area. **Electric current** is also carried by the flow of ions.

Light is composed of particles called photons. A **photon** is an elementary particle and is pure energy containing no mass.

Light can be produced by a variety of sources both electric and non electric. Non electric lights include gas lamps, kerosene lamps and candles. Electric lights include any light powered by electricity (including batteries).

An electrical **lighting fixture** is defined as an electrical device that is used to create artificial light. If you add a light source to the fixture, you have the beginnings of a luminaire. A **luminaire** is a light fixture with a light source, the housing to protect the light source, a reflector for directing the light (if needed), an **electrical ballast** (limits the amount of current in the electric circuit for this light if needed) and a connection to a power source. Light fixtures are classified by how the fixture is installed, the function of the light or the type of light.

> Albert Einstein observed that photoelectric light did not fit the classical wave theory of light. He realized that light energy had to be quantized, thus returning to the particle theory. The physics solution to this dilemma ultimately led to Einstein earning the 1921 Nobel Prize.

Several types of lights:

- Incandescent
- Light bulb
- Halogen light
- Heat lamp
- Fluorescent
- Compact fluorescent light (or lamp) (CFL)
- Light-emitting Diodes (LED)
- Neon lamp

Several types of light fixtures:

- Free standing table lamps, pole lamps, desk lamps, etc.
- Recessed lighting where the protective housing is embedded in a ceiling or wall with only the light source showing.
- Surface mounted lighting where the protective housing is exposed and not hidden, e.g. chandelier.
- Sconces which are usually mounted on walls and provide either up or down lights (illuminates artwork, provides lighting to hallways, etc.).
- Indirect lighting where light reflects off the ceiling for general illumination.
- Cove lighting where the light is in a long box against a wall in the ceiling.
- Track lighting where individual light fixtures can be positioned anywhere along a ceiling mounted track. The track provides the electric power to the fixtures.
- Outdoor lighting contained in enclosures that protect the light source from nature (wind, rain, heat, snow, etc.).
- Portable battery operated lights (glued or mounted) in closets, cabinets, etc. that provide light where there are no electrical outlets. Their self contained enclosures contain batteries.
- High bay and low bay lighting which is generally used to light manufacturing facilities.
- Strip lights or industrial lights which consist of long lines of fluorescent lamps also used to light manufacturing facilities and warehouses.

Emergency lighting and EXIT lights that are connected to batteries or to backup power that kicks in if there is a power failure. Emergency lighting and critical electrical systems are often connected to a UPS System. A **UPS System** is an uninterruptable power supply that provides enough power to service critical electrical systems that usually include emergency and exit lighting.

Several types of light fixtures controls:

- Light switches are on/off switches which close/open the circuit turning a light on or off. Light switches usually work on all types of light fixtures.

- Dimmer switches adjust light levels. The original dimmer switch increased the resistance of a variable resistor in the circuit, when a lower light level was desired. Unfortunately this solution consumed excess energy in the form of heat given off at the increased resistance. Since most households operate on alternating current, a new and improved method is to shut off the power each time the current cycles in a different direction for a set amount of time. The amount of shut off time is changed by the corresponding setting of the dimmer switch. The cycle is too fast for the human eye to actually see the light go on and off (in the US 120 times a second since the current cycles 60 times a second). The effect is an apparent dimming of the light when the shut off time is increased for each cycle. In reality the light does not actually go dimmer, it goes on and off too fast to see and just appears to be dimmer. Dimmer switches work best on incandescent lights. Dimmer switches shorten the life of most fluorescent lights.

- Occupancy sensors detect the presence of animate objects and turn on lights when something is detected. Sensors can be infrared (detect heat), ultrasonic (detect physical movement) and/or acoustic (detect sound).

- Timers turn on/off lights when a preset time is reached. Electronic timer switches have an internal open/close circuit switch which requires setting the time to close the circuit (turning on a light) and another time setting to open the circuit (turning the light off).

- A touch switch is an electronic device that enables the control of a circuit by touching a sensor.

Incandescent Lights

A little history

In the late 1700s and early 1800s, a large number of inventors were trying to create a means by which electric current would provide light. Thomas Edison is credited with inventing the first practical electric light. Thomas Edison had his own glass blowing operation and a number of assistants to create the bulbs he used for testing.

Thomas Edison tested thousands of materials including numerous plant fibers as well as metals in his search for a filament that would support the flow of electrons.

He built his first high resistance, incandescent electric light by passing electricity through a thin platinum filament in a glass vacuum bulb. The vacuum in the bulb delayed the filament from melting from the heat given off by the electric current. The lamp only burned for a

> Thomas Edison has been credited with inventing the electric light bulb. However, he was not the first. Joseph Swan, a British inventor, working at the same time as Edison, obtained the first patent for the same light bulb in Britain one year prior to Edison's patent date.

few short hours. Further testing by Edison and his team led to the use of carbonized cotton thread filaments. This provided a light equivalent to a 16-watt bulb and lasted for 1,500 hours.

Today, the filament in an incandescent light bulb is usually made of a long, thin length of tungsten metal. For a 60-watt bulb, the tungsten filament would be about 6.5 feet long but only 0.001 inch thick. The metal is coiled to fit inside the bulb. At this time, the standard incandescent light bulb is still very common in households.

How does a standard incandescent light bulb work?

Two metal wires go up into a glass bulb. Inside the bulb, the two wires are connected by a filament. The glass bulb is filled with an inert gas.

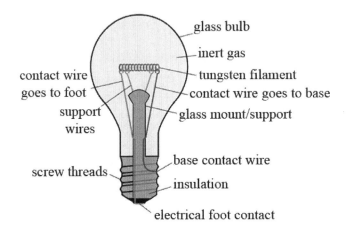

Drawing of a typical incandescent light bulb – cutaway showing wires and filament

Each wire from the bulb is connected to the wire in the electrical circuit. A power supply is connected to the circuit. Electric current flows up one wire through the filament and back down the other wire, completing the circuit.

When power is supplied to the circuit, the electrons move along the wire and through the filament. The electrons bump into the atoms in the filament causing movement of the atoms in the filament. The electrons in the filament atoms are not readily given up like in the wire (which is a conductor), but they are agitated sufficiently to orbit in a higher orbiting level around their nucleus. However, they are each still bound to its nucleus. The pull of the nucleus causes the electrons to fall back into their normal orbit. When they fall back into their orbit, they release the extra energy in the form of photons. These photons provide light.

To turn off the light when it is not needed, open the circuit. A light fixture control switch introduced into the electrical circuit will provide the means of shutting off the light. Any of the light fixture control switches (on/off switch, dimmer switch, occupancy sensor, timer or touch switch) can be used to open/close the circuit and result in turning off or turning on the light.

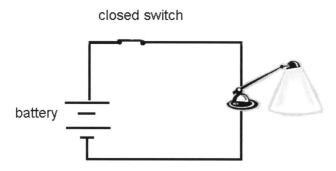

Drawing of a closed circuit with a light, power source and switch

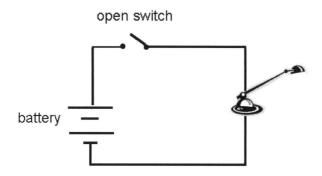

Drawing of a open circuit with a light, power source and switch

Disconnecting the circuit from its power source would also stop the flow of electrons and turn off the light.

Normal incandescent lights are not very efficient because they give off a significant amount of infrared heat when lit. Actually, they tend to give off more heat than light. This heat is wasted energy. If you look at the inside of the light bulb, you should be able to see a fine wire connecting the two metal wires coming up from the base of the bulb. Generally, after about 1,000 hours of normal use, the tungsten in the filament evaporates and deposits on the glass. A thin spot in the filament causes the filament to break and the bulb "burns out." When this fine wire is broken, the bulb will not light and the "burnt" out bulb and is of no further use.

Other Incandescent Lights

A **halogen light** also uses a tungsten filament. However, the filament in a halogen light is enclosed in a small quartz container. The gas inside the quartz container is halogen. When the tungsten vaporizes in the heat given off when current flows through the filament, the halogen gas combines with tungsten vapor. If the temperature is high enough, the halogen gas will combine with tungsten atoms as they evaporate and redeposit them on the filament.

When this occurs, this recycling process lengthens the life of the filament and consequently the halogen light. Due to the proximity of the filament to the container, a glass container would melt from the heat given off by the filament. Therefore, quartz is used as a container because it can stand higher temperatures than glass.

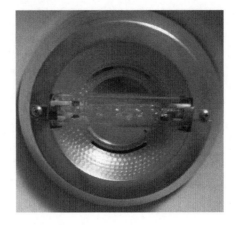

Picture of a halogen bulb in a lamp

Heat lamps take advantage of the fact that incandescent lights are inefficient and give off infrared heat. Heat lamps are used in restaurants to keep food warm and in certain greenhouses to nurture plants. Also, heat lamps are therapeutic and are used on humans to relax sore and strained muscles.

Picture of a heat lamp

Fluorescent Lights

A little history

A lot of people contributed to the development of fluorescent lights. Here are just a few of the contributors.

1857: French physicist Alexandre E. Becquerel investigated the phenomena of fluorescence. He experimented with coating electric discharge tubes with luminescent materials.

1896: Thomas Edison patented a fluorescent lamp that used x-rays to excite the phosphor. Edison was more interested in incandescent lighting and never sold this invention.

Mid-19th century, German physicist Julius Plucker and glassblower Heinrich Geissler found they could make light by passing an electric current through a glass tube containing small amounts of a gas.

1901: American Peter Cooper Hewitt patented the first mercury vapor lamp, a prototype of today's modern fluorescent light.

1902: Peter Cooper Hewitt formed Cooper Hewitt Electric Company to produce the first commercial Mercury lamps. His company was backed with money from George Westinghouse. His low-pressure mercury lights lacking red, made people look like corpses. In spite of this default, the mercury lights were successful because they put out many times more light than the carbon filaments of that era.

1927: Edmund Germer, Friedrich Meyer and Hans Spanner patented an experimental fluorescent lamp.

1938: George Inman, Richard Thayer and a team of General Electric scientists designed and sold the first practical and viable fluorescent light. General Electric bought the patent rights to Edmund Germer's earlier patent for $180,000 in order to strengthen its position against other companies working on the same application. Some individuals had already applied for patents, but Germer's German patent was already in place.

Fluorescent Lights

A typical fluorescent lamp is a sealed glass tube that contains a small bit of mercury and an inert gas (typically argon) kept under very low pressure. The inside of the glass is coated with a phosphor powder. There are two electrodes (one on each end), which are wired to an electric circuit.

Picture of a typical fluorescent light bulb

In 1923 Georges Claude and his French Company Claude Neon introduced the first commercial neon signs in the US. Earle Anthony, a Packard car dealer, purchased two signs reading "Packard." People would stop and stare at the "liquid fire" signs which were visible even in daylight.

When a fluorescent light is turned on, the circuit is closed and current flows throughout the circuit to the electrodes. When the current flows between the electrodes, the voltage drop across the electrodes is large. This forces electrons from the electrodes to migrate through the inert gas from one electrode at one end of the glass tube to the other electrode at the other end of the glass tube.

Mercury is liquid at room temperature. Heat converts the mercury in the tube from a liquid to a gas. Mercury has two valence electrons in its outer orbit. When electrons hit the gaseous mercury atoms, the electrons are bumped up to higher energy levels. The electrons release photons when they return to their original energy level. Since the mercury atoms release light photons in the ultraviolet wavelength range, the human eye does not see them. In order to create light with this process, the glass tube is given a phosphor powder coating.

When the ultraviolent photon from the mercury vapor hits a phosphor atom, the ultraviolet photon's energy is given to one of the phosphor's valence electrons. The phosphor electron is bumped to a higher energy level and the phosphor atom heats up. When the phosphor electron falls back to its normal level, it releases energy in the form of another photon. This photon has less energy than the original photon, because some energy was lost as heat, but it does give off **white light** that is visible to the human eye. This is why phosphors are substances that give off visible light when they are exposed to ultraviolet light.

Note: The disposal of fluorescent light bulbs is critical since the light bulb contains mercury. Care should also be taken if a bulb is broken. Local regulations apply to the disposal of products containing the toxic material mercury.

Neon lights were the first commercially successful fluorescents and are still primarily used for advertising. When a current is passed through a tube filled with neon gas, the result is a red glow. Neon is a noble gas. The **noble gases** are a group of chemical elements that are all odorless and colorless. Their outer shell valence electrons are full and the gases are considered inert with no negative or positive charge. Although other colors are generated by putting an electric current through a glass tube containing different noble gases, they are all called "neon" lights.

A neon sign

Light-Emitting Diodes (LED) Lights

A little history

Mid-1920s: Russian Oleg Vladimirovich Losev created the first LED.

1955: Rubin Braunstein of the Radio Corporation of America observed infrared emission generated by simple diode structures using gallium antimonide (GaSb), gallium arsenide (GaAs), indium phosphide (InP) and silicon-germanium, (SiGe) alloys.

1961: Experimenters Bob Biard and Gary Pittman of Texas Instruments found that gallium arsenide (GaAs) emitted infrared radiation when electric current was applied and received the patent for the infrared LED.

1962: Nick Holonyak, Jr. of General Electric Company developed the first practical visible-spectrum (red) LED. Holonyak is credited as the "father of the light-emitting diode."

1972: M. George Craford, a former graduate student of Holonyak, invented the first yellow LED and improved the brightness of red and red-orange LEDs by a factor of ten.

1976: T.P. Pearsall created the first high-brightness, high efficiency LEDs for optical fiber telecommunications by inventing new semiconductor materials specifically adapted to optical fiber transmission wavelengths.

1968: The Monsanto Corporation mass produced visible LEDs using gallium arsenide phosphide (GaAsP) for red LEDs to be used as indicators (the little red lights on electric appliances). Hewlett Packard (HP) used the GaAsP LEDs manufactured by The Monsanto Corporation for the alphanumeric displays in their early handheld calculators.

Light-Emitting Diodes (LED) Lights

Light emitting diodes, commonly called LEDs, are very tiny light bulbs that fit easily into an electric circuit. They differ from incandescent lights because they do not have a filament and don't give off a lot of heat.

LEDs are made of a semiconductor material typically aluminum-gallium-arsenide (AlGaAs). If the LED contained only AlGaAs, there would be no free electrons to conduct electric current because all of the atoms bond perfectly to their neighbors. So, doping is done to the semiconductor. Doping is the process of adding impurities to the semiconductor. The doped material has additional atoms changing the balance by adding free electrons or creating holes where electrons can go. Either of these additions modifies the semiconductor to make it into a better electrical current conductor.

When extra electrons are put into a semiconductor, the semiconductor is negatively charged and called N-type material. It should come as no surprise that a semiconductor with holes added to it is positively charged and called P-Type material.

Free electrons move from a positive area to a negative area in N-type material. In P-type material, the electrons jump from hole to hole, effectively moving from a negatively charged area to a positively charged area. Thus the holes in a P-type material "appear" to move from a positively charged area to a negatively charged area.

A diode consists of a piece of N-type material bonded to a piece of P-type material with electrodes at each end. Electricity can only flow in one direction in this configuration. When no voltage is applied, the free electrons from the N-type material fill the holes from the P-type material along the border between them. This creates what is called a depletion Zone. In the depletion zone, the semiconductor material returns to its original insulating state with all of the holes filled with the free electrons. Charge cannot flow. The depletion zone creates a barrier between the N-type piece and the P-type piece.

As long as the depletion zone exists, no current can flow through the diode.

To eliminate the depletion zone, the N-type piece is connected to the negative terminal of an electric source and the P-type piece is connected to the positive terminal. The electrons in the N-type material are repelled by the negative electrode. The holes in the P-type material "move" the other way.

When sufficient energy is applied increasing the voltage difference between the electrodes, the electrons and "holes" begin moving freely. The depletion zone will disappear and the charge will move across the diode. Note: if the energy source is accidentally hooked up backwards, instead achieving energy flow, the depletion zone will increase in size creating a larger barrier.

Photons are released as a result of the moving electrons and photons give off light.

Light emitting diodes (LEDs) are used in many places in electronics. LEDs form the numbers on digital clocks, the light on electrical appliances that says "I'm live," the newer form of Christmas lights and many other applications.

Differences Between Incandescent, Fluorescent Lights, and Light Emitting Diodes

Many compact fluorescent lights (CFLs) are designed to replace incandescent lights and can fit into existing light fixtures used by incandescent lights.

Conventional incandescent light bulbs do emit ultraviolent light but none of it is converted to visible light. That is the reason that a lot of energy used to power an incandescent light is lost as heat and wasted. Since a fluorescent light converts the invisible ultraviolet light to visible light, it is more efficient in the use of energy.

In an incandescent light bulb, the electric current flows through a solid conductor (wire). The electrical charge is carried by negatively charged free electrons (former valence electrons) jumping from atom to atom and moving from a negatively charged area to a positively charge area.

In a fluorescent light, the current goes through gas in a tube. In a gas, the electrical charge is carried by free electrons or ions moving independently from a negatively charged area to a positively charged area.

Compact fluorescent lights rated to give the same amount of visible light as incandescent lights generally use less energy, last longer, but have a higher purchase price. These tradeoffs of lower energy consumption, longer life and higher cost as well as a more complicated disposal process due to small amounts of mercury have to be balanced.

Light emitting diodes (LEDs) do not have filaments like incandescent lights or gases like fluorescent lights. Like fluorescent lights, LEDs generate more light and less heat than incandescent lights.

Observations

Lighting can be very complicated in many applications. The wrong kind of light in the wrong place can cause eye strain and headaches both in homes and in industry. Ergonomics is an extremely important aspect of lighting in order to maximize efficiency, comfort and well being.

Lighting can be extremely expensive. Studies on cost savings have shown that even though new lighting installed in older buildings is very costly, the savings in energy usually pay for the installation in a very reasonable period of time.

In new buildings, the time spent analyzing and designing the appropriate lighting for both function and aesthetics is well worth the effort. Lighting installations should look good and work well while taking advantage of the newer less energy consuming lights. There is an entire industry built around lighting.

Concept Reinforcement

1. Which of the following generates the most heat?

    - incandescent light
    - fluorescent light
    - light emitting diode

2. What is the difference between an ion and an electron?

3. What type of light in electrical appliances indicates whether the device is plugged into a live electrical outlet?

4. Is the incandescent light or the fluorescent light a more energy efficient device?

Section 3.3 – How Things Work – Portable Lights

Section Objective

- Explain how portable lights work

Background Information and a Quick Review

An **electric circuit** in a portable light consists of an energy source, a conductor to carry the electric current (flow of electrons) and a light source. Portable light electric circuits are direct current (current flows in one direction only).

A **battery** is an energy source. It has two terminals, one positive and one negative. When the terminals are connected to an electric circuit, the voltage is the force that pushes the electrons through the conductor. The amount of voltage is measured in volts. When the batteries are connected in series, the total voltage is the sum of the individual voltages.

Valence electrons are the electrons in the outer most orbit of an atom. They have a negative value. Their movement from atom to atom creates current. The amount of current is the number of these electrons flowing in a path in a conductor and is measured in amperes (amps).

A **series circuit** is where the current flows only in one path. A **parallel circuit** is where the current flows in more than one path.

A **light source** consumes energy and gives off light and heat. The amount of light and heat depends on the type of handheld device. The purpose of a portable light is to provide light. Any heat given off is considered energy wasted.

A typical portable light consists of the following components (the actual list depends on the design as some designs may not need a reflector or a lens):

- Case that contains the entire device
- External on/off switch that turns the light on/off
- Light source (incandescent, fluorescent, light-emitting diode (LED))
- Reflector that redirects the light rays from the light source, creating a steady beam of light
- Lens to protect the light source
- Energy source (battery or batteries)
- Conductor that connects the energy source to the light source and the on/off switch

Several types of portable lights:

- Standard portable battery operated incandescent flashlight
- Portable battery operated LED flashlight
- Head lamp (portable battery operated light)
- Clip mounted portable battery operated light
- Shake flashlight
- Portable lantern
- Novelty portable lights
- Portable fluorescent lamp

The energy source depends on the type of light source used. Several types of energy sources are:

- Alkaline batteries
- Lithium batteries
- Rechargeable batteries
- Button cells
- Solar powered energy sources
- Mechanically powered energy sources

A Little History

Some sources report that Joshua Lionel Cowen devised a lighting fixture for potted plants that consisted of a light bulb in a metal tube powered by a dry cell battery. Conrad Hubert, a Russian immigrant, supposedly purchased Cowen's company and patents in 1898 and renamed Cowen's company the American Electrical Novelty & Manufacturing company. Joshua Lionel Cowen was the manufacturer and founder of Lionel toy trains and was quite successful in marketing his toy trains.

Conrad Hubert first ran a cigar store, then a restaurant, boarding house and a novelty shop. Whether or not it was Cowen's company he purchased, Hubert was clever enough to see the potential in a hand-held battery-powered torch. He hired David Misell, a British man working in Hubert's New York shop in 1898, to produce a tubular portable flashlight.

Originally, the battery was only able to sustain light for a few seconds, hence the name "flash" light. Hubert donated several flashlights to the New York City police. Even though the flashlights using zinc-carbon batteries did not sustain light for very long, they were favorably received.

Misell assigned the flashlight's patent rights to Hubert's new business American Electrical Novelty and Manufacturing Company. Hubert marketed the flashlights under the Ever Ready brand name. Eventually his company became The American Ever Ready Company. As you might suspect, Conrad Hubert was quite successful and made a lot of money.

Standard Handheld Flashlight

A standard handheld flashlight is powered by batteries. It has an on/off switch, a light bulb, a protective lens covering the light bulb, metal strips that conduct current from the batteries through the closed switch to the light bulb and back, a reflector that surrounds the bulb and reflects the light around the light bulb to one direction, all in a protective case.

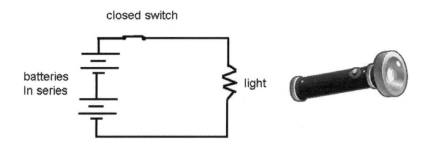

Picture of a typical flashlight showing the path for current to flow

The typical flashlight has more than one battery connected in series. The negative side of one battery rests on top of a small conductive spring in the bottom of the flashlight. The small spring is connected to one of the conductive strips of material. The other battery rests on top of the first battery. The batteries are connected in series so that the flow of electrons runs between the positive and negative electrodes (terminals) of the batteries. Connections to the rest of the circuit (light and on/off switch) are made by strips of conductive material that carry current.

The contact strip from the bottom of the battery (which rests on the spring) runs inside the flashlight case until it makes contact with one side of the switch. There is another flat contact strip on the other side of the switch which runs to the light. A flat contact strip runs from the light and makes contact with the positive electrode of the top battery.

When the on/off switch is set to the on condition, the switch closes the circuit by connecting the two strips of conductive material.

Electrons flow from the series of batteries along the contact strip, through the now closed switch and into the light. The electrons pass through the metal contacts to the light bulb in the flashlight. A thin wire in the light bulb contains the element "tungsten", which glows when in contact with electricity, producing a visible light. This light reflects off of the reflector that is positioned around the lamp. The reflector redirects the light rays from the lamp, passing through a glass lens and creating a steady beam of light. This is the light you see coming from the flashlight. The glass lens shapes the light into a broad beam and also provides a protective covering for the light.

When the flashlight switch is set to the OFF position, the two contact strips are physically moved apart and the path for electrons to flow is broken. The flow of electric current stops and the light goes out.

Portable Battery Operated LED Light

When a light emitting diodes (LEDs) replaces the incandescent light in a battery operated portable light, the major difference is in the light itself. The circuit with the on/off switch, the batteries in series (if more than one is required) and the conductive contact strips are similar to the standard incandescent flashlight.

The LED gives off light when an electric current pass through it, just as the light does in the incandescent flashlight. The electroluminescence (color, etc.) of the LED light depends upon the semiconducting material used to construct the diode. While LED lights are typically very small, manufacturers add special lenses to help with the reflection. The result is a powerful beam for the small size of the LED.

LED lights are usually embedded in a high-impact epoxy which makes them very durable. They have no loose or moving parts and produce little or no heat. This makes them very useful in moving vehicles such as cars, boats, and planes.

Head Lamp (Portable Battery Operated Light)

Lights can be mounted on hard hats, attached permanently to head straps or clipped on to a head strap. The variety of lamps is quite extensive. They can be incandescent lights, LEDs, or fluorescents lights. The only unique thing about head mounted lights is that they are designed for hands free work. Some of the lights swivel and move up and down so they can be directed where light is needed. They come in handy when working on automobile engines, working in a mine, or anywhere that over head light may not be accessible. It is much easier to work when the light is in line with where a person's eyes are directed. They are also used by cave explorers.

Clip Mounted Portable Battery Operated Light

Clip mounted portable lights are a convenience. Like the head mounted lights, the variety of choices is extensive and can be incandescent lights, LEDs, or fluorescent lights. The hardware to attach a light to a clip is simply that – hardware. Then the light is located where it will provide the desired lighting. Actually, any means of mounting a light will work. Duck tape has been used to tape a lamp to a particular location.

Shake Flashlight

The shake flashlight is different than simple battery operated flashlights. The shake flashlight has a light bulb. It requires electric current to provide energy to make it light. It can have an incandescent bulb with a small wire filament just like the traditional flashlights or it can use the newer LEDs just like LED flashlights. The difference is in how energy is provided to the light.

Most shake flashlights have a generator and a capacitor in place of a battery. The generator consists of a hollow copper coil with a permanent magnet inside of it. This magnet is in a track, which allows it to slide up and down when the flashlight is shaken. When the flashlight is shaken, the magnet passes through the copper coil repeatedly and the magnetic field creates a flow of current.

The flow of current powers up the capacitor and the capacitor provides current to the light. As long as there is current flowing through the capacitor, the light will be on. When the light starts to dim, shaking the flashlight will cause the magnet to pass through the copper coil again generating more current. The flashlight should shine brighter.

Capacitors do not store as much energy as batteries and will not hold their charge as long as batteries. But since batteries work on chemical processes, they will eventually wear out. The capacitor is basically two sheets of foil with a piece of paper between the two pieces of foil and is not likely to wear out.

Since LEDs do not lose as much energy to heat as incandescent lights, the LEDs can run for much longer on fewer shakes.

The disadvantage of using a shake flashlight is that it works off a magnetic charge. The directions contain a warning to not set the light close to computers, magnetic storage media and other sensitive electronic devices. Someone who wears a pacemaker should probably not use a shake light.

Portable Lantern

Portable lanterns function very much like regular flashlights except they are bigger. The lights are larger and the batteries are larger.

Portable lantern

Novelty Portable Lights

Novelty portable lights are just that – novelties. Who says that novelties can't be useful? A cute light may be just the thing for a child to hang onto. The novelty lamps essentially work the same as regular flashlights.

A portable frog light

Another fast growing group of novelty lights are small animal or gadget lights attached to a key chain. These lights give off very small amounts of light, but they can be very useful. For example, they give off just enough light to see the key lock when trying to open a locked door in the dark. They are a nice thing to have when the door is outside and you are standing in the freezing rain or snow trying to get into your home.

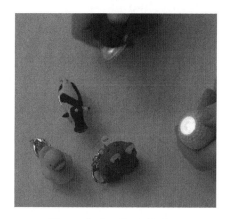

Key chain flashlights

Portable Fluorescent Lamp

These lamps work the same as any other portable lamp. They have bulbs, batteries, on/off switches, and internal wiring to connect the batteries to the switches and to the light bulb. They give off more light than heat and last longer than the incandescent variety. These lamps can actually replace reading lights during a power outage. They can also be very useful on camping trips where there is no electricity.

A fluorescent lamp and side by side components

Summary

All portable lights are in an enclosed container with a light, an on/off switch, conductors to complete the circuit and an energy source. The energy source is usually a battery, but there are now hand held lights that can be charged through mechanical motion (your labor builds the energy). Multiple batteries are in series to provide more voltage to the light. Portable lights are simple in design but valuable in many ways.

Observations

Portable lights are a marvelous convenience. We can take our light with us when we go camping. They are useful in lighting up any dark place that is not wired for electricity. They are great for looking under our beds (shows up the dust bunnies, too). They help us find things we have lost that we otherwise can't see. As a child I huddled under the covers with a flashlight so I could continue to read after my parents said "Lights out! It's time to sleep." They can be fun. Who of you has put a flashlight under your chin in a dark room to scare a younger brother or sister?

However, portable lights have many important uses such as automobile headlights. Without headlights, driving cars at night would be a little scary and very dangerous. Minors used head lights when they were working in the mines. Rescue workers use portable lights when they enter disaster areas which no longer have electricity. Portable lights come in very handy during power outages.

Safety checklist:

- Make sure there is a working portable light in the family car
- Make sure there are working portable lights strategically placed throughout your home (basement, bedroom, family room, garage, etc.)
- Keep spare batteries on hand for each portable light
- Keep spare lights for incandescent and fluorescent portable lights
- Periodically check that the portable lights are in working order (and especially before a storm)
- Keep portable lights and spare batteries in all emergency kits (make sure your family has an emergency kit)
- It is a good idea to keep one of the mechanically charged lights in the emergency kit. While they can be a lot of work to keep going, they will provide light in an emergency for much longer periods of time than their battery operated counterparts.

Note: if a light will not used for a long time, the batteries can be removed and kept with the light so they will be available when needed.

Concept Reinforcement

1. Are the batteries in a standard incandescent flashlight connected in series or in parallel?

2. What are the 7 components of a typical incandescent flashlight?

3. True or False. On/Off switches are not required in a portable light.

4. True or False. It doesn't matter which way batteries are installed in a portable light.

Section 3.4 – How Things Work – Car Batteries

Section Objective

- Explain how a car battery works

A Quick Review

A **car battery** is an energy source. It has two terminals, one positive and one negative. The terminals are part of an electric circuit. The amount of voltage produced by the battery is measured in volts.

The amount of **current** is the number of electrons flowing in an electric circuit along a conductor and is measured in amperes (amps). Electrons have a negative value and move from the negative terminal of a battery through an electric circuit back to the positive terminal of a battery.

In a **series circuit** the current flows in only one path.

A Little History

Did you know that the first recorded self-propelled road vehicle was a steam powered military tractor invented in 1769 by French engineer Nicolas Joseph Cugnot?

Cugnot also built a 4-passenger steam powered 3-wheeler in **1770**. Over the years, numerous other steam powered vehicular inventions followed. The most successful steam powered vehicles were locomotives.

Robert Anderson of Scotland invented the first electric carriage in the **1830s**. Professor Stratingh of The Netherlands and his assistant Christopher Becker built a small electric car in **1835**.

In **1842,** American Thomas Davenport and Scotsman Robert Davidson each built electric road vehicles using the first non-rechargeable electric cell batteries.

Better batteries led the way to better electric cars. France and Great Britain led the development of electric cars in the **late 1800s**.

Thomas Hugh Parker, son of the founder of the Electric Construction Company (ECC) was a brilliant innovator, and claimed to have had a battery powered car running as early as **1884**. In **1891,** Americans, A.L.Ryker and William Morrison built a six- passenger wagon.

Americans were excited about electric cars. In **1897,** the first commercial electric cars were built by the Electric Carriage and Wagon Company of Philadelphia and a fleet of electric taxis hit the streets of New York City.

Then in **1899,** Camille Jénatzy, a Belgian, built an electric car that set a world record for land speed at 68 mph. Wow!

Although electric, steam and gasoline vehicles were all available in America, in **1899 and 1900,** electric cars outsold all other types of cars. The electric car was the choice of many for several reasons. Electric cars did not require hand cranking like gasoline vehicles. They did not need a gear shifter. Electric cars didn't stink, make noise and vibrate like gasoline cars. Steam cars were not popular because they took too long to start up on cold mornings and needed frequent water stops.

With the head start that electric cars had at the turn of the century, why did they die out? Electric cars were the toys of rich people. Electric car prices continued to rise. Then Henry Ford began mass production of the internal combustion engine. He made cars available and affordable to the masses for the first time. Americans built roads between cities and the shorter-range electric cars could not compete with the internal combustion engine. The oil fields of Texas provided affordable gasoline. By **1935,** the electric cars became museum pieces.

It is interesting to note that that as early as **1916,** Woods invented a hybrid car that had both an internal combustion engine and an electric motor.

The internal combustion engine became "king." Gas and internal combustion cars were affordable. Americans took to the highways. Very little work was done to develop and mass-produce electric or hybrid cars between **1920 and 1965.**

As early as **1960,** the need for solving exhaust emissions and reducing dependency on imported foreign crude oil was identified. In 1966, in an effort to reduce air pollution, the U.S. Congress introduced the first bills recommending the use of electric vehicles. **1970** saw the beginning of the up and down gasoline shortages. Efforts to build economical alternate vehicles began in earnest.

A number of companies and entrepreneurs worked on electric vehicles and hybrids. Some were successful and some were dismal failures. In 1977, Toyota sold nearly 18,000 Prius hybrid cars in Japan. All-electric cars did not fare well and most development was dropped by 1999.

In **1999,** the Honda Insight was the first hybrid car to be mass released in the United States. EPA mileage ratings of 61 mpg in the city and 70 mpg on the highway were very exciting. Then in **2000,** Toyota released the Prius, the first four-door sedan hybrid available in the United States. When Toyota released the Prius II in 2004, the U.S. demand was so high that people waited up to six months to purchase this hot new hybrid.

In **2008,** the American car industry was in financial trouble. Gas prices were all over the chart. Hybrid cars and internal combustion cars were both available. Interest in all-electric cars continued, especially for city use due to their low pollutant emissions.

Car Batteries

What is the Purpose of a Car Battery?

A car battery is the energy source that provides the temporary surge of amps needed to start the motor of a car. When the car is running, the internal charging system (the alternator and voltage regulator) continually recharge the battery and power the electrical systems of the car. The electrical systems include headlights, windshield wipers, fans, heaters, brake lights, internal lights, the radio, the CD player, etc. Basically, the car won't start without a battery and some of the electronics are critical. The car may run without the headlights, but it would be tough to see at night.

Lead-Acid Batteries for Internal Combustion Engines

The style used to start most internal combustion engines is a rechargeable lead acid battery. The lead acid battery is the oldest style of rechargeable battery and has been in use since the early 1900s.

A car battery is called an SLI (starting – lights-ignition) battery. It is designed to deliver quick bursts of energy so that it can start the car engine. A high power output from a car battery is necessary to start the car's engine.

A normal car battery provides 12.6 volts. The standard battery is composed of six separate units called cells. The cells are connected in series to form a "battery of cells" called a battery. Since each cell produces about 2.1 volts, the total output of the battery is 6 times 2.1 volts. This provides a total draw of 12.6 volts. The high power output of a car battery is necessary to supply the high current draw needed to start the engine. The power of the battery is created through a chemical reaction.

Note: a car battery just off a charger and fully charged may register 13.8 volts.

Sulfuric Acid (H_2SO_4) is a strong mineral acid. It is soluble in water (H_2O) at all concentrations. The percentage of acid in water that is used in the lead acid batteries is 33.5%. That means that 66.5% is water (H_2O) and 33.5 % is sulfuric acid (H_2SO_4). The common name for the combination of water and sulfuric acid at this percentage is "battery acid." The battery acid solutions are electrolytes.

There are lead plates inside each battery cell. These lead plates are immersed in the battery acid (H_2SO_4 and H_2O). One set of plates is lead. The other set of plates is coated with lead dioxide. Lead plates and sulphuric acid generate lead sulphate as the battery provides energy (discharges). The reverse chemical process occurs (lead sulphate is destroyed) when the battery is regenerated.

The lead plate reacts with the ionized sulphuric acid to produce lead sulphate, hydrogen ions in solution and two excess electrons.

$Pb + HSO_4^- \rightarrow PbSO_4 + H^+ + 2e^-$ designated as the anode

The lead oxide reacts with the ionized sulphuric acid, the available hydrogen ions and the donated excess electrons to produce lead sulphate and water.

$PbO_2 + HSO_4^- + 3H^+ + 2e^- \rightarrow PbSO_4 + 2H_2O$ designated as the cathode

Car batteries

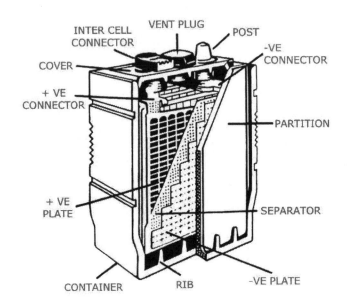

Cross section of car battery

The lead compounds are insoluble and stay attached to the plate so there is never any significant amount of lead in solution. The water and sulfuric acid are completely soluble. These properties are an important part of why the cell is rechargeable.

When the solution interacts with the lead plates, the chemical interaction creates voltage. This voltage is then released through the batteries positive terminal (red) and returned to the negative terminal (black) when the circuit is closed. The circuit is closed when the key is turned in the switch to start the car.

When the circuit is closed (the key is turned in the ignition), the two terminals of the battery are connected and excess electrons from the anode will travel through the circuit as current and return to the cathode.

As the battery discharges it creates two chemical reactions, one at the anode that ends up with an excess of electrons, and one at the cathode that ends up short electrons. As the battery discharges, both plates build up $PbSO_4$ and water in the cells.

When the circuit is closed and the voltage from the battery is released, the current flows to the ignition to start the engine. Current also flows through the electrical systems in the car powering the headlights, the radio, electronic displays, etc.

Lead-Acid Battery Ratings

Batteries have standards. These standards are used to rate the output and capacity of an SLI (starting-lights-ignition) battery.

- Cold cranking amps (CCA) is the number of amps a battery can deliver at 0°F for 30 seconds before the voltage out drops below 7.2 volts. In cold weather, a higher CCA rating is better.

- Cranking amps (CA) is the number of amps a battery can deliver at 32°F. Also called marine cranking amps (MCA).

- Reserve capacity (RC) is the number of minutes a fully charged battery at 80°F will discharge 25 amps before the voltage out drops below 10.5 volts.

Recharging a Lead-Acid Car Battery

Recharging a battery is very simple. The battery provides sufficient energy to start the motor of the car. Electrons flow from the negative side of the battery through the car electrical circuit and returns to the positive side of the battery. To recharge the battery, the flow of electrons must be reversed to restore the battery to its charged state.

As the battery discharges, the voltage produced is decreased. While the engine is running, the alternators main function is to convert power from the gasoline engine that drives the car along the road, to charge the battery that powers the car's electrical system.

Note: A "gel cell" is usually a lead-acid battery that has something in the sulphuric acid solution to make it less liquid and more like a gel. Gel cells require slower charging than standard lead-acid batteries because they have more trouble dissipating heat.

How to Deal with a Dead Lead-Acid Car Battery

When is the Battery Dead?

When the key is turned in the ignition and the engine does not crank, the first suspect is the battery. First, verify that the battery is the problem.

Symptoms of a dead battery (when the key is turned to start):

- No sound

- A single click

- A rapid series of clicks

- A slow cranking sound

The most common cause for any of these symptoms is a weak or dead battery or a poor connection to the battery.

A simple test to check the battery

Turn on the interiors light and try to start the car. The battery is probably ok if the interior light is bright when you turn it on and doesn't dim when you turn the key to start the engine. This means you have other problems.

If the interior light is dim or goes dim when the key is turned, the battery or battery connection is suspect.

Common reasons for a dead battery or no battery charge

- Leaving the headlights on
- Leaving a reading light or courtesy light on
- A defective switch for internal or trunk light that leaves light on
- The battery is old and has lost its ability to maintain a full charge (cold weather forces batteries to work harder and shorten their life)
- Battery cables are loose or corroded and do not carry current to the starter

Options for recharging a dead battery

- Call for help
- Use an external battery charger (make sure the charger will charge the battery at the proper rate)
- Find a car with a good battery and battery cables

The acid-lead batteries do require care in handling

Battery acid is very corrosive. Wash any skin that comes into contact with lots of water. If any gets your eyes, you should flush them thoroughly with water and see a doctor IMMEDIATELY.

- Wear safety goggles
- Wear special gloves
- Wear old clothes with long sleeves and long pants

A premixed solution of baking soda and water on hand can help neutralize spilled acid and minimize the damage. Do not get any baking soda inside the battery or you will destroy the battery. If you have a spill, dump the baking soda -water solution on the spill. The soda will fizz until the acid is neutralized. Then water can be used to cleanup the spill.

Do not touch the positive terminal on the battery. The positive terminal is insulated and goes to all the components in the car that require power. DO NOT CLOSE THE CIRCUIT WITH YOUR BODY!

When charging, a lead-acid battery emits hydrogen gas. To avoid an explosion, DO NOT SMOKE OR HAVE ANY OPEN FLAMES IN THE AREA!

NEVER CHARGE A FROZEN BATTERY. It may explode because of the gas trapped in the ice. Remove the battery and thaw it out before attempting to charge it.

How to recharge a car battery using jumper cables

Heed the warnings in the previous section.

Wearing an old pair of gloves, check the battery connections (and do not touch the terminals with anything metal)

Check the battery itself to make sure it doesn't have any leaks or cracks. If it does, skip the rest of this list and call a tow truck.

Once you are protected, check to see if the connections are loose on each battery terminal. Be careful not to damage the battery post. If the connection is loose, see if tightening the connection will solve the problem.

A car battery has two terminals either on top or on one side of the battery. They should be marked with a (+) for positive and a (–) for negative. They are also probably color coded with red for positive (+) for and black for negative (-). The negative terminal is directly connected to the metal body of the car as well as the metal engine block. This is the Ground. The positive terminal is insulated and goes to all of the electrical components in the car.

Find a friend with a car and a good battery and jumper cables. You want to make sure the jumper cables are heavy cable with insulated wire at least 6 gauge (4 gauge is better – thicker wire).

Check the owner's manual for both cars. On some cars, the manufacturer does not recommend jump starting. Other manufacturers specify certain procedures to follow, such as removing a specific fuse. The manufacturers procedures written in the owner's manual should always be followed.

If there are no "don't dos" in the owner's manual, when jump starting a dead battery from another car, follow this procedure:

Put the front of the cars close together WITHOUT TOUCHING!

Shut off the engines in both cars and open both of their hoods.

Wear eye protection! THIS IS IMPORTANT!

The cables must be connected in this exact order:

- Connect one end of the (+) positive cable to the (+) terminal on the **dead** battery.

- Connect the other end of the (+) positive cable to the (+) terminal on the **good** battery.

- Connect one end of the (-) negative cable to the (-) terminal on the **good** battery.

- Connect other end of the (-) negative cable to the engine block of the dead car or a metal bracket that is directly attached to the engine. DO NOT ATTACH THIS CABLE TO THE DEAD BATTERY! If you get a big spark, something is wrong and you should not attempt the jump start.

Next shut off all electrical equipment on the good car. The electrical equipment in the dead battery car should also be turned off.

NO SMOKING OR OPEN FLAMES in the area.

Let the good car's engine run for a couple of minutes before attempting to start the dead car.

After a couple of minutes, get in the dead car and attempt to crank it. If it sounds like it wants to start but won't quite turn over, give it some more time and try again.

If the car is still hard to start, first disconnect the negative cable from the bad car, then check to make sure that there is a good solid connection at each of the remaining cable clamps, then reconnect the negative clamp on the disabled car's engine block and try again.

Once the dead car has started and is running for a few minutes, it is time to remove the jumper cables. This must be done in reverse order:

- Disconnect the (-) cable from the engine block on the former **dead** car.

- Disconnect the (-) negative cable from the (-) negative terminal on the **good** battery.

- Disconnect the (+) positive cable from the (+) positive terminal on the **good** battery.

- Disconnect the (+) positive cable from the (+) positive terminal on the former **dead** battery.

Keep the car with the former **dead** battery running until the battery can either be recharged by driving or with a battery charger.

Hybrid Car Batteries

The hybrid car has both a gasoline engine and an electric motor. It uses the gasoline engine under normal driving conditions, but the electric motor kicks in extra power when it is needed. When the hybrid needs additional energy, it can draw energy from the batteries to accelerate the car. Acting as a generator, it can slow the car down and return energy to the batteries when acceleration is no longer needed.

The batteries in a hybrid car are the energy storage device for the electric motor. The gasoline fuel tank provides the source of power for the gasoline engine, but it can't put gas back in the tank once it has been used. The electric motor on a hybrid car can put energy back into the batteries as well as draw energy from them.

The key to the efficiency of the hybrid car is that the gasoline engine is not sized for maximum performance. That extra boost of power is provided by the electric motor. In an internal combustion only car, the car is sized larger to handle those situations which require extra power. Most of the time that extra power is not needed.

The hybrid car sometimes shuts off the gasoline engine when it is idling. It can do this because it has an electric motor and batteries. When cruising down the highway, the gas engine is doing all of the work. When the hybrid slows down by hitting the brakes for letting up on the gas, the electric motor kicks in to generate electricity to charge the batteries. The hybrid doesn't have to be plugged in because it charges its own batteries.

The battery pack in a hybrid vehicle contains hundreds of cells. Instead of battery acid that is used in lead-acid batteries, the NiMH battery cells contain a chemical mixture called nickel metal hydride (NiMH). The NiMH batteries come with a long time warranty of 8 years or 80,000 miles. The typical lead-acid battery has a life span of about 4 years give or take a few depending on weather and usage.

The negative electrode reaction in a NiMH cell is

$H_2O + M + e^- \leftrightarrow OH^- + MH$ where the metal "M" is an intermetallic compound

The electrode in the NiMH cell is charged when the reaction goes to the right and discharged when it goes to the left. On the positive electrode, nickel oxyhydroxide (NiOOH) is formed:

$Ni(OH)_2 + OH^- \leftrightarrow NiO(OH) + H_2OP + e^-$

Lithium-ion battery

Electric Car Batteries

An electric car is powered by an electric motor. It does not contain a gasoline engine. The electric motor is nearly silent. The electric motor gets its power from a controller. The controller gets its power from an array of rechargeable batteries.

Many types of batteries have been used in electric cars including lead-acid, nickel cadmium (NiCd), nickel metal hydride, lithium ion, Li-ion polymer, zinc-air and molten salt.

The battery is the weak point in electric cars. They are still expensive and require frequent charging. Fuel cells are also being investigated. The future is wide open.

Summary

The lead-acid battery in internal combustion engines has a short life time and disposal issues. Under certain circumstances, it can be dangerous to handle. Lead-acid batteries have been known to explode. The spillage of sulphuric acid can be very damaging.

A gasoline engine has fuel lines, exhaust pipes, coolant hoses and intake manifolds and looks like a plumbing project. An electric car is a wiring project. A hybrid is both. As gasoline prices fluctuate and environmental concerns increase, the pressure is on to find alternatives to the internal combustion engine.

Concept Reinforcement

1. A lead-acid battery usually has many battery cells?

2. If each battery cell is 2.1 volts, what is the calculated voltage of a lead-acid battery?

3. What device closes the circuit when starting a car?

4. In a lead-acid battery, where does the energy come from?

Section 3.5 – How Things Work – Water Pump in an Internal Combustion Engine of an Automobile

Section Objective

- Explain how the water pump in a car engine works

Background Information

What is a pump? It is a device that moves liquid material from one place to another. Liquids can be defined as anything that flows such as gases, water, water mixtures, slurries, or other viscous materials.

What makes a pump work? Unless the pump is a manual pump, electric current provides the energy to power the movement of the pump and the pump moves the material.

What are some of the common uses of pumps? Think of the Alaskan pipeline that pumps oil across many miles. Think of a water pump that provides running water to your home. Think of a pump that pumps seawater out of a boat. Think of the water pump in an internal combustion engine car that pumps coolant (water plus antifreeze) to cool the engine.

A Little History

A 'shadoof' is a form of a manual pump. It was first identified in 2000 BC in ancient Mesopotamia. A long pole is suspended on a frame above a water hole. The pole is suspended on the frame about 1/5 from the end of the pole. A bucket or container is on the long end of the pole and a weight or counterbalance is on the short end. When the container is half full, the weight and container are balanced. The same amount of effort is required to lower the empty container into the water and raise the full container. This device is surprisingly efficient and is still used in some areas of the world today.

Manual pumps have a long and colorful history. An example of a common manual pump is the hand bicycle pump. When electricity arrived on the scene, energy sources provided the "muscle" required to move the parts inside a pump that moved material. There are now thousands of pumps ranging from the manual 'shadoof' pumps to electric pumps. Pumps can move anything that flows.

Types of Pumps

Pumps are chosen primarily for a specific function. They are the force that moves liquid, viscous, or gaseous material from one place to another. Just like voltage pushes electrons through a conductor, pumps push liquid or viscous materials through pipes. In many texts, the movement of electrons being pushed by an energy source is explained by using the analogy of a pump pushing gallons of water through a pipe.

Pumps come in many types and many sizes and are made by many different manufacturers. Pumps are also made out of different materials depending upon the application. Some materials are very corrosive. For an application that requires the movement of corrosive material, the pump must be made of material that can withstand contact with the corrosive material. If the pump is made of material that cannot withstand the corrosive material, it can be destroyed by the material it is trying to pump.

Two of the more common types of pumps are positive displacement pumps and centrifugal pumps.

Positive displacement pumps contain rotating (pumping) elements such as gears, rotors, screws, or vanes. The pump creates a space between its pumping elements. The liquid being pumped moves into and is temporarily trapped in the space between the pumping elements. As the pumping elements rotate, the size of the space is decreased and the liquid is forced out of the pump. In other words, a discrete amount of liquid is trapped in a chamber which is alternately filled with liquid from the inlet and emptied through the discharge.

This section will address the centrifugal pump. The centrifugal pump in its simplest form consists of an impeller rotating inside a casing. The impeller consists of a number of blades mounted on a shaft. The shaft extends out of the casing. Power is applied to the shaft rotating the impeller inside the stationary casing. The revolving of the impeller blades causes a decrease in pressure at the entrance of the impeller. Liquid enters the casing near the rotating axis of the impeller. Liquid is forced outward along the impeller blades at an increasing velocity and exits the pump at a higher pressure. Centrifugal force is the force that sends the coolant outward where it exits the pump at a higher pressure. The Law of Centrifugal Force is the basis for the success of the centrifugal pump. Centrifugal force is an outward force based on the rotating impeller. While centrifugal force is considered a pseudo-force, never the less, the effects are quite real.

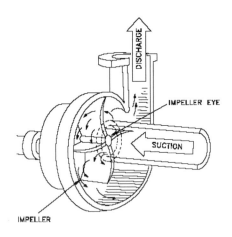

Diagram of a centrifugal pump

A typical water pump in an internal combustion engine in an automobile is a centrifugal pump.

Audi water pump
courtesy of Bill Linesfelser

Purpose of water pump in an internal combustion engine

When an internal combustion engine "burns" fuel, it gives off lots of heat. The engine must be cooled otherwise it will overheat and cease to work. The water pump is like a heart for the car's cooling system. Like your heart beats and circulates blood throughout your brain and body, the water pump circulates engine coolant (water plus antifreeze) through hoses around the engine into the radiator and back to the pump.

The purpose of the radiator is to remove heat from the coolant as the coolant flows through the radiator. The radiator is a type of heat exchanger. The coolant flows into the radiator and flows through a lot of tubes mounted in parallel in the radiator before leaving the radiator. The tubes inside the radiator are cooled by outside air blowing throughout the radiator as the car is moving. The heat from the coolant (which is on the inside of the tubes) is transferred to the air (which is on the outside of the tubes) blown through the radiator.

There are two systems working simultaneously. One system pumps the coolant around the engine, through the hoses, through the radiator and back to the pump. That system cools the engine. The other system uses air to remove heat from the coolant inside the radiator. In the first system the coolant absorbs heat by cooling the engine. The second system cools the coolant.

In the second system the coolant gives off heat that is absorbed by air blowing throughout the radiator. Both systems must work together to keep the engine cool. When the car is moving at normal speeds during normal temperatures, the air flow in the radiator is sufficient to cool the coolant. However, when the car is idling, moving too slowly through traffic or the ambient temperature is very high, the temperature of the air is too high to cool the coolant in the radiator with normal air flow. When the temperature is high enough, the temperature sensor triggers a fan. The fan is turned on and forces the air flow through the radiator to help cool the coolant. The "cooled" coolant exits the radiator and returns to the pump to continue its job of cooling the engine. The hot air exits the radiator.

What powers the water pump in an internal combustion engine?

The modern automobile's electrical system supplies energy for a number of functions. For example, energy is required to start the engine and keep it running even though the internal combustion engine runs on fuel. Energy is needed for the fuel pump, the water pump, cooling fan, headlights, turn signals, inside lights, radio, CD player, power windows, inside heaters and air conditioners just to name a few. There are electrical sensors that monitor what is going on in the car and electronic displays that tell the driver what is happening.

This lesson will address the water pump. An energy source is required to turn the impeller blades inside the centrifugal pump. Typically a belt is connected to the crankshaft of the engine and this belt drives the impeller shaft in the centrifugal pump. On some cars the overhead cam engines drive the water pump with a timing belt. In any case, a belt of some type is connected to the shaft of the impeller. When energy is applied, the belt turns the impeller. As the impeller turns, coolant is pumped out of the water pump through the hoses around the engine into the radiator and back to the pump.

Water pump installed in Audi
courtesy of Chris Neal

Location of timing belt that drives water pump
courtesy of Chris Neal

If the belt connected to the crankshaft turns the impeller, what powers the crankshaft?

A simplified description of how to start the engine and activate the water (coolant) pump:

- The key in the car's ignition switch closes the circuit allowing current to flow from the battery to the starter via the starter solenoid switch.

- The starter makes the engine "turn over" (engine starts running).

  - The starter starts an electric motor called a starter drive

  - The starter drive turns a ring gear

  - The ring gear spins a flywheel that is attached to the crankshaft

  - The crankshaft is attached to the pistons

- Electric current stops flowing to the starter once the engine starts and flows through the electrical system through the alternator.

- The battery and alternator are in a closed circuit. The alternator recharges the battery while the car engine is running.

- The electrical circuit also powers all the electronics in the car. If extra current is needed occasionally for the electronics during normal operation of the car, the battery can provide the additional current as needed (too much extra current can drain the battery – if the alternator cannot recharge the battery to compensate for the excess usage, the electrical systems could fail).

- Once the engine starts running, the fuel pump pumps fuel from the fuel tank into the intake manifold and then the combustion chamber. Fuel is mixed with air.

- The fuel/air mixture flows into the cylinders in the engine.

- In each cylinder is a piston which traps the fuel/air mixture and compresses it into a very small space in the cylinder. The pistons are attached to the crankshaft which starts turning when the engine is started.

- The ignition system provides high voltage to the spark plugs causing a spark. The spark from the spark plug ignites the fuel/air mixture which burns and expands the mixture.

- The expansion of the fuel/air mixture forces the piston back down with more power than when it moved up to compress the fuel/air mixture.

- A connecting rod is attached to the bottom of the piston.

- The connecting rod is also attached to the crankshaft. As the piston rods move up and down, they keep the crankshaft turning. The crankshaft turns around and around like a wheel. The crankshaft is physically connected to the drive train and is ultimately responsible for turning the car wheels and making the car go.

- Also attached to the crankshaft is a belt. This belt is attached to the impeller shaft of the water pump. As the crankshaft goes around, not only does it make the car move, it also forces the impeller blades to rotate and pump the coolant.

Summary

The water pump is in reality a coolant pump which pumps coolant through the engine and keeps the engine from overheating. The cooling is necessary because when the fuel/air mixture burns and expands, it reaches very high temperatures. **The water pump is indirectly powered by the electrical system in the car. It is actually operated by the belt attached to the crankshaft of the engine.**

The crankshaft is initially turned by energy from the battery when the ignition system is activated by the key. The crankshaft starts the up and down motion of the pistons. Once the pistons are set in motion and the fuel/air mixture is ignited by sparks from the spark plugs, the pistons will continue their up and down motion based on energy derived from the burning/expansion of the fuel/air mixture.

The ignition switch remains as a closed switch in the electric circuit which is tied into the battery and alternator.

Once the pistons operate on the fuel/air mixture, their continued up and down motion takes over the turning of the crankshaft. The pistons will continue to operate (and the engine will run) as long as there is fuel pumped from the fuel tank, the ignition switch remains in the on position providing a closed circuit (the spark plugs need that spark to ignite the fuel), the cooling system cools the engine and nothing major goes wrong with the car.

Concept Reinforcement

1. What is the primary purpose of a pump?

2. True or False. All pumps require electric current to operate.

3. What is the energy source required to start an internal combustion engine?

4. What is the function of an ignition switch once the engine has been started?

Section 3.6 – How Things Work – Ceiling Fans

Section Objective

- Explain how a ceiling fan works

A Little History

Before air conditioning became so common, people used manual fans to stir the air creating a wind chill effect. While fans do not actually lower the temperature, they do create air flow which creates a cooling effect by evaporating perspiration on the skin.

For those of you who live in cold windy areas, fans have the same effect as a wind chill. The effect of the wind creates a perception of a temperature lower than the actual temperature.

The first ceiling fans in the 1800s were not powered by electricity. They were driven by a system of belts powered by water turbines.

Philip Diehl engineered the electric motor used in the first electric Singer sewing machines. In 1882, he modified the sewing machine electric motor to operate a ceiling-mounted fan. The fan was a self-contained unit mounted on the ceiling and drawing electrical power to operate. This turned out to be a "hot" project that generated a lot of competition.

Diehl added a light fixture as part of the ceiling fan. This was a cool idea since now the fan would not have to displace overhead lighting. The combination is very similar to what is in operation today.

Ceiling fans went from two blades to four blades in the early 1900s. By 1920, ceiling fans were very common in the United States and internationally. Once the United States entered the Depression followed by World War II, ceiling fans had become almost non-existent in the United States. However, ceiling fans remained very popular in warm climate countries.

Starting in 1950s and 1960s, ceiling fans started slowly to come back in use in the United States. In the 1970s energy concerns triggered a rebirth in the use of ceiling fans when it was determine that they consumed less energy than air conditioning units.

Photos of a running ceiling fan

The primary purpose of a ceiling fan is to move air gently keeping the occupants of a room comfortable.

Background Information

- An **electric circuit** contains a power source (V), a conductor to carry electrons and a resistance (R).

  - An **energy source (V)** pushes electrons along a conductor.

  - A **resistance (R)** consumes energy.

  - **Electric current (I)** is the flow of electrons.

  - An **electric switch** provides a means to close or open an electric circuit (start or stop an electrical device).

- **Alternating Current** is common in households and business. Current reverses direction periodically.

- **Direct Current** flows only in one direction.

What is a ceiling fan and how does it operate?

A ceiling fan is a fan that is mounted on the ceiling. It usually has 5 blades, a light and a self-contained motor. Wired into a normal household circuit, it operates on alternating current. Depending on the model purchased, the ceiling fan can have multiple speeds, reverse capability and a light. Reverse capability means the motor contains a switch which reverses the rotation of the fan.

The ceiling fan can be purchased to operate using pull chains. One chain turns the light on/off. Another chain turns the motor on/off and changes the speed. The two chains are independent circuits. The light can be on while the fan is off and the fan can run without the light. The second chain when pulled goes through a cycle of speeds. The first pull selects one speed. The second pull selects another speed. It is common to have three speeds, but not all models have multiple speeds. Again it depends on the model that is purchased. Some fans just turn on or off and have only one speed. Some fans do not have a light fixture.

On some models, a remote wall switch can operate the fan. In the example below, the energy source to the fan is turned on by the bottom rocker switch. The light is turned on by the top rocker switch. The fan is turned on by the middle rocker switch. Pressing and holding for a few seconds changes the speed of rotation of the fan. Pressing the light switch and the fan switch together reverses the rotation of the fan. The fan will coast to a stop. Then pressing the fan switch starts the fan rotating in the reverse direction.

A wall switch that operates a fan with light and reverse capability

How does a ceiling fan function?

When the fan is turned on and the blades are rotating clockwise, the warm air is drawn up towards the ceiling. While ceiling fans do not actually change the temperature in a room, they create a wind chill affect which makes the room feel cooler. Just like a nice breeze outdoors makes a person feel cooler, the air movement caused by the ceiling fan pulls perspiration away from the body allowing a person's own cooling system to work more effectively.

Multiple speed ceiling fan with light and reverse speed capability

When the ceiling fan turns clockwise, the high edge of each blade is catching air and the low edge is deflecting air. The air flow is up towards the ceiling. If you stand directly under the fan, you do not feel the air flow.

When the rotation of the fan is reversed (counterclockwise), it's just the opposite. The low edge now catches air and the high edge deflects air. The air flows downward. If you stand directly under the fan when it is moving counterclockwise, you can feel the air blowing directly down on you.

If you run the ceiling fan counter clockwise in the summer, you can expect some savings on your air conditioning costs. Fans are cheaper to run than air conditioners. By running ceiling fans with air conditioning, the temperature of the air conditioner can be set higher by several degrees without changing the level of comfort. You can even reduce winter heating costs by running ceiling fans clockwise.

Fan blows air down.

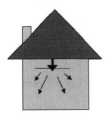

In summer, the downward movement of air creates a cooling "wind chill" effect. A room can actually feel several degrees cooler, without setting the thermostat lower.

Fan draws cool air up.

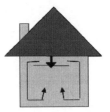

In winter, hot air rises to the ceiling while cool air settles to the floor. Trapped against the ceiling, the warm air is wasted. Circulating the fan in reverse, moves war air down from the ceiling into the living area and draws cool air up.

What are the components of a ceiling fan?

The actual components depend upon the manufacturer and the design. Every ceiling fan must have the following:

- An electric motor that will turn the blades. Typically this is a direct drive spinning motor that runs on alternating current (AC).

- One to six blades (or paddles) usually made of wood, metal, or plastic mounted on the motor (above, below or on the side of the motor).

- Blade irons, blade brackets, blade arms, blade holders or flanges which are metal arms that connect the blades to the motor. A new design called a rotor is an alternative to blade irons. Ron Rezek patented the design in 1991 which consists of a one-piece die cast rotor that bolts directly on the motor. The rotor holds the all the blades of the fan. The advantage of the rotor is that it eliminates most balance problems.

- A means to mount the fan to the ceiling such as a ball and socket, a J-hook (sometimes called a claw-hook), or a special kit from the manufacturer.

- An on/off switch for the fan's motor

Optional components include the following:

- A downrod which is a metal pipe that suspends the fan from the ceiling.

- A decorative motor housing for the motor

- A switch housing which provides a place for a light kit

- Blade badges, which are decorative ornaments on the visible underside of the fan blades. They make the fans look attractive while hiding screws used to attach the blades to the blade irons.

- Switches that permit changing of the speed of the fan and the direction of rotation.
- Lamps and their on/off switches.

Compare impact of typical ceiling fan on typical air conditioning unit

To calculate the energy consumed by a ceiling fan:
energy consumed = energy x time

- Wattage × hours/day / 1000 = daily kilowatt-hour (kWh) consumption

Assume the fan is rated at 85 watts.

In the summer, assume the fan is running at its highest speed and using 85 watts

- 85 watts × 24 hours /1000 = 2.04 kWh energy consumed in one day

A window air conditioner rated at 3000 watts

In the summer, assume the air conditioner running 70% of the time would provide a comfortable temperature of 70 Degree F.

- 3000 watts × 16.8 hours /1000 = 50.4 kWh energy consumed in one day

Assumptions:

By running the ceiling fan, the air conditioner can be set to run at 80 Degree F with basically the same comfort level.

With the temperature set for 80 Degree F, the air conditioner would only have to run 60% of the time.

Note: these are assumptions made to show an example of the impact of using ceiling fans to reduce overall energy costs. Outside temperature and the size of the room are not being considered in this example.

To show the reduction in energy consumption by reducing the amount of time the air conditioner runs:

- 3000 watts × 14.4 hours/1000 = 43.2 kWh energy consumed by air conditioning running at 60% of the day.

The savings in energy consumption of the air conditioner by changing the air conditioner's thermostat from 70 Degree F to 60 Degree F is:

- 50.4 kWh – 43.2 kWh = 7.2 kWh energy savings in one day

However, remember that to achieve that energy savings, the ceiling fan must run 24-hours/day. This brings the daily consumption of energy up to:

- 43.2 kWh (air conditioner) + 2.04 kWh (ceiling fan) = 45.24 kWh total energy consumed

In this hypothetical example, the total energy of the air conditioner (A/C) by itself would have been 50.4 kWh per day. By running the ceiling fan, the air conditioner can be set to run at a higher temperature and still maintain the desired comfort level. The combined energy of the ceiling fan and the air conditioner at the higher temperature is 45.24 kWh per day. This provides an overall savings of:

- 50.4 kWh (A/C alone) − 45.24 kWh (A/C + fan) = 5.16 kWh savings per day.

Assume that the utility company charges $0.09 for each kWh consumed. Also, assume that this took place where the air conditioner and fan were both used for a period of 90 days. Under these conditions, the cost savings would be:

- $0.09/kWh × 5.16 kWh × 90 days = $41.796

While this is a hypothetical example, the actual savings can be calculated based on the devices used to cool (their ratings) and the amount of time they actually ran (time when they were cycling) times the cost of the kWh. The cost of kWh can be obtained from the local utility bill or by calling up the local utility company and asking their price.

Summary

In summary, ceiling fans can effectively reduce both cooling (summer) and heating (winter) costs by improving the circulation of air. In the summer, the ceiling fan blows air downward providing a breeze that acts like a wind chill increasing the comfort level even though the actual temperature is not affected by the ceiling fan.

In the winter the direction of the rotation of the ceiling fan is reversed. The fan circulates the warm air that rises to the ceiling back down into the living area providing even heating and comfort.

Concept Reinforcement

1. True or False. The primary purpose of the ceiling fan is to circulate air.

2. True of False. Air conditioners and ceiling fans should not be run at the same time.

3. Assume that the air conditioning unit was not energy efficient and when it ran by itself without the fan it ran 24 hours /day. What would be the daily energy consumption of the 3000 watt air conditioner running 24 hours/day?

4. The ceiling fan has a chandelier added to it and someone left the lights turned on. The energy consumption for the ceiling fan is now 90 watts instead of 85 watts. What is the daily energy consumption of the ceiling fan still running 24 hours /day?

Section 3.7 – How Things Work – Portable Electric Heaters

Section Objective

- Explain how a portable electric heater works

Purpose

The primary purpose of a heater is to warm an area making people comfortable. Heat can be provided by a furnace using air ducts to heat an entire house/building. Outdoor heaters can provide warmth by generating blasts of heat in a specific area. Indoor single unit heaters can reduce the overall need to heat an entire house or building by keeping specific areas warmer.

This lesson will focus on portable electric heaters. The purpose of this lesson is to understand how the generation of electricity can produce heat in portable electric heaters. Since improper use of portable heaters can be hazardous, we will also study safe operating practices and review the features that are built-in to safeguard use of the unit.

A Little History

Thomas Ahearn was one of Canada's groundbreaking businessmen and inventors in the early age of electricity. He invented the first electric car heater in 1890. Thomas Ahearn founded the Ottawa Electric Railway Company, which provided electric street car services in Ottawa. These were the first streetcars with electric heaters.

A Quick Review

Current is the flow of electrons along a conductor.

A **conductor** is made up of material that easily gives up valence electrons allowing electrons to flow.

A **resistance** is any material that impedes the flow of electrons.

Heat is generated when electrons come upon a resistance that impedes their flow.

Heating capacity is measured in **British Thermal Units (BTUs).**

BTUs measure heat (energy).

Watts measure power (energy per unit time).

1 watt = 3.412 BTU/hour.

Wattage ratings of heaters can be converted to BTUs consumed per hour by multiplying the number of watts by 3.413 (the number of BTUs/hour equaling one watt).

Examples of Heaters

The following are some examples of the various types of heaters available. The purpose is to acquaint you with several of the types of devices available and some of their operating characteristics, not to promote any specific unit.

Type of Heater	Example	Description
Ceramic Portable Space Heater		Portable, convection heater designed to heat small areas.
Fan Heater		Portable, convection heater uses a fan to pass air over a heating element and blow warm air out of the heater. Some fan heaters can be noisy.
Infrared Heater		Quiet and portable, these heaters don't use as much energy as convection or fan heaters.
Infrared Heater		Top and sides are cool to touch. Very stable unit. Electromagnetic waves transfer thermal energy.
Oil-filled Heater		Electricity warms up a special type of heat-conserving oil that is sealed inside the coils of the radiator. Since the oil is only warmed, never burned, and never needs to be replaced or refilled, it makes oil-filled radiator heaters very efficient to operate. The oil absorbs heat and radiates this heat into the surrounding area. Considered safe and efficient.
Simulated Fireplace Heater		Safer than wood burning, these portable heaters operate exclusively on electricity.
Baseboard Heater		Baseboard heaters are not portable, but they are safe and easy to install. Electric resistance is used to warm the air that passes through. Because they are on the floor, warm air flows out the top (warm air always rises) and cold air is pulled into the bottom to be heated.

Kerosene Heater		Convection heater circulates warm air upward and outward in all directions. Not to be used in enclosed space because burning kerosene consumes oxygen. May be illegal is some areas. Also available as radiant heater. Kerosene is highly flammable. Built-in battery operated lighting device is preferable to using matches to light the wick.
Propane Heater		Convection, radiant and infrared models available.
Natural Gas Heater		Not portable because it must be piped to natural gas and vented to outside. This unit uses natural gas to warm soapstone, which provides radiant heat. While this model does not have a blower to move warm air, most gas heaters do. Usually has electric lighting device.
Wood Heater		Burns wood or coal. Made out of cast iron, heats by radiation. No electricity involved. Not legal in states of California and Washington. Should not be used in closed area.

How Heaters Work

Convection Heaters

Convection heaters work by converting electricity into heat. An electric current flows through a resistor (element) that provides resistance to the flow of electrons. The element gets hot. A fan blows air across the heating element. The air is warmed and then blown out to heat the air in a room.

Convection heaters with a reflector and one or more fans provide uniform warmth in a room and are very effective. Some models with fans oscillate helping to move air around the room.

Convection heaters warm the air in the room. The warm air makes people feel warmer.

Ceramic Heaters

Ceramic heaters are a type of convection heater. They are small portable electric heaters that contain a ceramic disk heating element and aluminum baffling. The ceramic plates are the resistance in the electrical circuit. Electric current flows through the unit and finds a resistance in the ceramic plates. The ceramic plates become hot. The aluminum absorbs the heat from the ceramic plates. The air surrounding the aluminum gets hot. A fan blows the hot air into the room.

Ceramic heaters are small and lightweight. Easy to move around, they provide localized heating. They are great under a desk for keeping feet warm.

While they heat up quickly, the outside of the units are made of material that does not absorb heat and is cool to touch.

Infrared Heaters

There are many types of infrared heaters. Some contain metal-sheathed tubular heaters, and some quartz tubes or lamps.

Infrared heating is a direct form of energy transfer that does not rely upon conduction through air.

There are three primary types of infrared defined by wavelength

- near-infrared (wavelength = 0.076 to 1.5 microns)
- mid-infrared (wavelength = 1.5 to 5.6 microns)
- far-infrared (wavelength = 5.6 to 1000 microns)

Far-infrared is an invisible band of light that warms objects without heating the air between the source and the object. If you stand in the sunlight on a cool day, the warmth you feel is the effect of the far-infrared wavelengths penetrating your body. Because far-infrared waves have a similar resonance with water, the water molecules in the human body absorb the far-infrared waves directly and transform the radiant energy into thermal energy or heat.

Infrared heaters use far-infrared wavelengths to transfer thermal energy to objects and occupants in a room.

Example: One style of infrared heater uses a quartz bulb, which contains a filament. The filament inside the quartz bulb provides resistance. When electric current is passed through the quartz bulb, the metal filament in the bulb gets very hot. The quartz bulb is inside a stainless steel metal diffuser coil. When the filament gets hot, the diffuser coil gets hot. The infrared heater contains copper-infrared heat chambers. The heat chambers when heated by the quartz bulb wrapped in a stainless steel coil generate far-infrared waves. These far-infrared waves heat the objects and occupants in the room directly at the speed of light. The warmth when the heater is turned on is felt immediately.

When the infrared heater is shutoff, an internal fan runs to cool the heating elements. The unit should not be unplugged until the fan stops.

Why Use Portable Heaters

If central heating is too costly or the main heating source is inadequate, then smaller heating units can be utilized to raise the temperature of a smaller living or working area. Rather than heat an entire house or building, a subset of that space can be heated with independent units. This is sound economic sense only if the cost of operating the smaller units is less than the cost to raise the temperature and heat the entire building.

The good news about space heaters is that if they are confined to smaller living areas, they do not lose energy through ducts or to areas not in use. The choice of an energy efficient heater that is safely operated is an excellent solution for increasing the comfort level in a specific home or work area.

Safety Features

Portable electric space heaters increase comfort when used properly. When not used properly, they can be a serious fire hazard. Some of the safety features you should consider are:

- Automatic shutoff when unit overheats.
- Automatic shutoff when unit is tipped.
- Thermostat controls to shut heater off when temperature is reached.
- Tag or stamp of approval from an independent testing company such as Underwriters Laboratories.

Safe Operating Procedures

- Electric heaters should be plugged directly into the wall outlet. If an extension cord is necessary, it must be heavy duty. Check with the manufacturer to verify the gauge of wire necessary to insure safe operation of the unit.
- NEVER CUT OFF THE ROUND GROUNDING CONNECTOR TO MAKE IT FIT A TWO PRONG OUTLET! Make sure the plug fits properly in the electrical outlet and that it is not loose.
- Make sure the electrical outlet will provide sufficient power (wattage) to operate the electric heater. If too many electrical devices are connected to the circuit being used, it may trip and overload when the heater is turned on.
- NEVER HIDE CORDS UNDER CARPETS AND RUGS! This could damage the cord and cause overheating.
- Periodically inspect the cord and connector to insure they are in good condition.
- Make sure the heater is not next to anything that is flammable. Heaters should be 3 feet away from furniture or curtains.
- Do not put anything on top of the heater.

- Do not block air flow around heater.

- Place the heater on the floor and not on top of any piece of furniture. Besides, since heat rises, they work better when they are on the floor.

- Most heaters come with a tip over safety switch. If the heater is knocked over it should shut off immediately. Be aware that the surface of the heater may be so hot that contact with something flammable could cause a fire.

- Always turn off heater when leaving the room for any length of time.

How to Size the Space Heater

To obtain the most efficient usage from a portable eater, it should be sized properly. Some of the factors to be considered are the size of area to be heated, will it be indoors, in a garage or outside. If it is indoors, is the area insulated.

Calculate the area to be heated. Measure the width, depth and height of the room. Multiply the height of the area by the width and depth to obtain the total space to be heated.

Estimating is not an exact science. It is based on a combination of facts and experience in that field. It is important to **document all assumptions** made when providing any engineering estimate. For the engineering estimate to determine the BTUs required to heat a room that has 12 foot ceilings and is 10 feet wide and 12 feet deep, the following assumptions are made:

- The room is insulated.

- The room will be closed (windows and doors can be closed while the heater is in use).

- An electrical outlet is available that will accept the plug of the device (for example, in the United States, a three-hole plug will be required).

- The electrical outlet will provide sufficient power (wattage) to operate the electric heater. If too many electrical devices are connected to the circuit being used, it may trip and overload when the heater is turned on.

- The heater can be located close enough to the outlet so that an extension cord is not required. If an extension cord is required, it must be of sufficient gauge (gauge is the number of the conductor wire which is really the size of the conductor wire) to support the flow of current. Once the size of heater is determined, the appropriate size of wire (gauge) can be determined. The manufacturer can specify this.

Estimating BTUs:

Method 1

A rough calculation for determining the required heat output (number of BTUs) needed is to assume that for each cubic foot, you need 5 BTUs. If you multiply the total cubic feet of space to be heated by a factor of 5, this will give you a rough approximation of the number of BTUs required to heat the area.

Example: A room has 12-foot high ceilings and is 10 feet wide and 12 feet deep. How many BTUs would be required using this estimating process?

$Area_{total}$ = 12 ft × 10 ft × 12 ft

$Area_{total}$ = 1,440 cubic feet (ft^3)

To estimate the BTU requirements:

1,440 ft^3 × 5 BTUs/ft^3 = 7,200 BTUs

According to this method, you would need approximately 7,200 BTUs to heat a room of this size.

Method 2

A rough calculation for determining the required wattage is to calculate the square feet of floor space and multiply by 10 watts per square foot.

Note: 1 watt = 3.412 BTU/hour.

Example: A room has 12-foot high ceilings and is 10 feet wide and 12 feet deep. How many BTUs would be required using this estimating process? Since this estimating method only uses square feet of floor space, the ceiling height is not used in the calculation. If the ceiling is exceptionally high or very low, this will definitely make a difference in the results.

Floor area = 10 ft × 12 ft

Floor area = 120 ft^2

To obtain the wattage requirement:

120 ft^2 × 10 watts/ft^2 = 1200 watts

Using this method, the BTUs/hour would be:

1200 watts × 3.412 BTUs/hour/watt = 4094.4 BTUs/hour

Comparing Method 1 and Method 2

Comparing Method 1 and Method 2, the results are different. The difference between 7,200 BTUs and 4,094 BTUs may be significant. Look at the numbers in a different light. Look at the values in watts instead of BTUs.

7,200 BTUs/hr → 2,110 watts (remember to divide BTUs/hr by 3.412)

4,094 BTUs/hr → 1,200 watts

The values in watts don't appear to have as big a difference, but they are still different. Remember, these were only estimates using very rough calculation methods. When an engineer provides an estimate, the engineer uses the best technical data available and his/her personal experience. If the engineer does not have the personal experience to determine which estimate is reasonable, a good engineer would continue to investigate. Method 3 would provide another option.

Method 3

To estimate the required heat output needed, calculate the total area to be heated and contact several manufacturers of heaters. Since they should have done the research, they should be able to provide you with the required size of unit needed to heat the area using one of their products.

Energy Saving Tips

Small portable space heaters can be convenient and increase the comfort level in small living and work areas. Ways to keep the energy usage as low as possible include the following:

- Determine the area you wish to heat and purchase the proper size heater for that area.

- Select a heater that is thermostatically controlled to maintain the temperature. The unit should shut itself off when it gets too warm and turn itself back on when the temperature drops. This will prevent overheating.

- When leaving the area for any length of time, shut off the heater or lower its temperature setting until you return. Turning the heater on and off frequently is inefficient, so if you are just leaving the room for a few minutes, leave it on.

- If a timer is available, set it so that the unit will shut down after a certain amount of time. That way the unit will not continue to run overnight in the event someone leaves and forgets to turn it off. If you are still in the area when it shuts off, you can always turn it back on for another block of time.

- **Make sure that the heater selected is properly vented** and keep doors and windows closed in order to minimize heat loss.

Summary

Fueled heaters work well because they use little to no electricity, but these types of heaters cannot be used in areas without adequate ventilation. Portable electric space heaters work well in small closed areas. Electricity is available almost everywhere and electric heaters do not require any fuel to be stored.

Electric heaters are safe as long as safe operating procedures are used. Some are more efficient than others and there are many choices. Some are decorative and some are not.

Convection electric heaters use electricity to heat a resistor (element). The heat built up in the element is transferred into the area being heated. The method of transfer depends on the type of heater. Convection heaters use fans to blow hot air across the element. The air is warmed by the element and then blown into the room. Some units do not have a fan.

Infrared heaters use electromagnetic waves to bombard objects in the room thus heating them.

Heaters come with a variety of options, but some of the safety options are most important. Automatic shutoff if overheating is very important. Automatic shutoff that turns the power off if the unit is tipped is extremely important.

Heaters should not be placed near anything flammable. Electric cords and plugs should be periodically inspected for wear.

Electric heaters if used properly can provide temporary energy efficient solutions to cold.

Concept Reinforcement

1. True or False. Infrared heaters use near-infrared waves to heat objects.

2. Multiple Choice. Which of the following is a convection heater?
   A. ceramic heater
   B. radiant heater
   C. infrared heater

3. True or False. Electric heaters should never be turned off.

4. True or False. An estimate is always 100% accurate.

Section 3.8 – How Things Work – Microwave Ovens

Section Objective

- Explain how a microwave oven works

A Little History

During World War II, two scientists invented a magnetron tube that produced high frequency short waves called microwaves. Microwaves are reflected by most metals (they will not penetrate metal). This characteristic led to the invention of radar. The magnetrons were installed as part of Britain's radar system. This enabled the Britains to identify Nazi warplanes on bombing runs to their country.

In 1946, an engineer by the name of Dr. Percy LeBaron Spencer was working in his lab at Raytheon experimenting with a vacuum tube called a magnetron. When he reached into his pocket, he found that his candy bar had melted. This intrigued the good engineer so he exposed un-popped popcorn to the magnetron rays. Much to his delight and amazement, popcorn crackled and popped all over his lab. Now he was really interested in the side effects of this tube. He next exposed a raw egg to the magnetron rays. The egg exploded and a curious coworker ended up with yolk all over his face.

Dr. Percy LeBaron Spencer

Like most new inventions, the first microwave built to cook food was large and expensive in the beginning. By 1975 the sales of microwave ovens exceeded the sales of gas ranges in the United States.

What is the purpose of microwave ovens?

Microwave ovens are a modern convenience for heating and cooking food very quickly. Microwave ovens can be found in restaurants, manufacturing facilities, break rooms and most homes. The primary purpose of microwave ovens is to heat/reheat/defrost or cook food or beverages very quickly.

What are microwave ovens and how do they work?

Photo of a built-in microwave oven

Photo of a countertop microwave oven

Photo of the inside of a microwave oven

Microwaves are very high frequency radio waves that swing back and forth (change electromagnetic fields from positive to negative) at a frequency of 2 billion cycles per second. Microwave ovens use microwaves to produce exactly the right wavelength to excite water molecules. Excited water molecules move faster and increase in temperature. Most food contains a significant amount of water. When the water molecules heat up and rotate, they bump into other molecules inside the food. The process is like frictional heating. Microwave energy converts to heat energy by causing food molecules to move faster. This heats the food product. The larger the proportion of water in the food, the faster the food product will heat up.

Dancing water molecules?

Water is H_2O. A water molecule contains two hydrogen atoms and one oxygen atom. Water molecules are polar molecules. The electrons are shared so that one side of the molecule is positive and the other side is negative. The positive side of the water molecule has a positive field around it and it repels other positive fields while being attracted to negative fields. The negative side of the water molecule has a negative field around it and it repels other negative fields and is attracted to positive fields. Like charges repel and unlike charges attract.

Now if you add the electromagnetic fields produced by a microwave, the electric fields inside the microwave are constantly changing. The water molecules in the food or beverage react to the electromagnetic fields by rotating to minimize the force on both the positive and negative ends of the molecule. When the electromagnetic field changes from positive to negative, the water molecules rotate again (minimizing the force). The water molecules are doing a kind of dance. Each time the electromagnetic field changes, the water molecules dance away. As the water molecules dance, they increase in heat. They transfer this heat to the food molecules.

The frequency of microwave ovens is 2.45 gigahertz. This changing of the electric field from positive to negative about 2 billion times a second makes the water molecules dance.

The Magnetron

The magnetron tube is the guts of the microwave oven. In order to send out electromagnetic waves sufficient to excite the water molecules and heat food, a high voltage system is required. The microwave oven's transformer increases the household voltage of about 115 volts to approximately 3000 volts using a capacitor and a special diode. The magnetron tube inside the microwave oven converts the high voltage to electromagnetic waves.

A magnetron is made up of the following parts:

- Anode – an iron cylinder that controls the movement of the microwaves
- Cathode – a filament that emits electrons that travel back and forth to the anode
- Antenna – a probe that helps guide the energy emitted from the magnetron
- Magnetic field – created by strong magnets mounted around the magnetron

Electric current is provided to the cathode. Electrons move into the space between the cathode and anode. Since the anode is positive and both the cathode and the electrons are negative, the electrons follow the antenna towards the anode. The electrons encounter the magnetic field. The forces of the electric charge and the magnetic field cause the electrons to enter into an expanding circular orbit until the electrons finally reach the anode. The circular motion of the electrons produces the microwaves of energy.

What not to put in a microwave

The key to cooking products in a microwave is to use containers that permit microwaves to pass through and to make sure there is sufficient water in the food or beverage product to generate heat. Current studies recommend using glass containers to heat food (not plastic).

A microwave should never be turned on when it is empty.

Live creatures should not be put into microwave ovens. Remember that microwaves can boil water. Most creatures are around 90% water.

Microwaves will pass through glass and plastic containers. A good test to see if a dish is microwavable is to **fill it with water** and heat for 2 minutes. If the dish is hot, it should not be used in the microwave.

Microwaves bounce off metal. If you put a metal object inside the microwave oven and turn it on, the microwaves will bounce off the metal and bounce around inside the microwave oven. The microwave oven is basically a metal box and will contain the microwaves. The metal wiring in the glass window of the microwave oven permits light to pass through but keeps the microwaves from leaving the oven. There should be something inside the oven that can absorb the microwaves. The magnetron can be damaged if the microwaves are bounced back to the source (the magnetron). Sufficient bouncing microwaves could cause electrical sparks inside the oven. Besides, if food is put into a microwave in a metal container or aluminum foil, the microwaves will bounce off the metal and will not heat/cook the food.

Safety feature: The microwave oven will not operate with an open door. If your microwave does not shut off when the door is opened, it is defective and needs to be repaired or replaced.

How to calculate energy consumption of a microwave

Microwave cooking is different from conventional oven cooking. Conventional ovens heat and cook by conduction. They cook from the outside of the food product to the inside. Conventional ovens are usually set to high temperatures and create crusts on the outside of foods.

Microwaves cook from the inside out. Microwaves heat food by exciting molecules. Conventional ovens heat by conduction.

To calculate the energy consumed over a specified period of time, multiply the amount of energy (watts) times the time duration of the operation.

Energy consumed = energy × time

Wattage × Hours Used per Day ÷ 1000 = Daily Kilowatt-hour (kWh) consumption

1 kilowatt (kW) = 1,000 Watts

So, for a 600 watt microwave operating at full power for 10 minutes, the amount of energy consumed is:

600 watts × 10/60 hours = 100 watt-hours

100 watt-hours/1000 = 0.1 kWh

If the microwave is rated at 750, operating at full power for 5 minutes, the amount of energy consumed is:

750 watts × 5/60 hours = 62.5 watt-hours

62.5 watt-hours /1000 = 0.0625 kWh

Compare typical microwave to typical electric oven

To cook a frozen dinner entrée

Microwave directions – cook at full power for 7 minutes

Microwave rated at 1800 watts

1800 watts × 7/60 hours = 210 watt-hours

210 watt-hours/1000 = 0.21 kWh

Then cook at half power for 7 minutes

Microwave rated at 1800 watts at half power

900 watts × 7/60 hours = 135 watt-hours

135 watt-hours /1000 = 0.14 kWh

Total usage to cook dinner in microwave

0.21 kWh + 0.14 kWh = 0.35 kWh

If the electric utility company charges $0.09 per kWh, the cost to cook the dinner in a microwave would be

0.35 kWh × $0.09/kWh = $0.03

Conventional oven directions–cook at 350 Degree F for 55 minutes

Oven rated at 2100 watts for baking, operating for 55 minutes, the amount of energy consumed is

2100 watts × 55/60 hours = 1925 watt-hours

1925 watt-hours /1000 = 1.92 kWh

If the electric utility company charges $0.09 per kWh, the cost to cook that same dinner in a conventional baking oven would be

1.92 kWh × $0.09/kWh = $0.17

It costs $0.14 more to cook the meal in the oven instead of the microwave. Even though $0.14 doesn't sound like a lot of money, try cooking one meal a day in the microwave instead of the oven and the annual savings would be 365 × $0.14 = $51.10

Concept Reinforcement

1. What is the primary purpose of a microwave oven?

2. True or False. Microwaves use more energy than conventional ovens to cook food.

3. True or False. Metal containers are as good as glass for heating food in a microwave.

4. If a microwave is rated at 1000 watts, what would be the energy consumption (watt-hours) to heat a cup of water for 2 minutes?

Section 3.9 – How Things Work – Wireless Remote Controllers

Section Objectives

- Explain how a wireless remote control works

A Little History

1899: A Madison Square Gardens audience thought they were controlling a small boat with voice commands. But Nikola Tesla was listening and secretly sending signals to the boat using a remote controller with radio signals tuned to control the operation of the small boat. It's good to know that the great Nikola Tesla did have a sense of humor.

1906: Leonardo Torres Quevedo remotely guided a boat from the shore of Bilboa in the presence of the King and a great crowd. This demonstration of his invention Telekino (a robot) is considered the birth of remote control.

1904: Jack Kitchen, a Lancashire inventor, demonstrated a radio-controlled torpedo.

1920s: The navy used various radio-controlled ships for target practice.

1930s: The Soviet Red Army used remote controlled tanks in the Winter War against Finland and in World War II. Britain developed and used radio controlled aircraft for target practice.

1939: Philco's Mystery Control was the first wireless remote controller available to the public. It had radio station numbers on the face place instead of numbers and the radio could be "dialed" using the remote device from anywhere in the house. There is no record of the first "lost" remote controller.

1944: The Germans used radio controlled motor boats filled with explosives to attack enemy ships.

1950: Zenith Radio Corporation developed the first remote television controller. Unfortunately, it was a tripping hazard since it was physically connected to the TV by a wire. It did have a neat name. It was called "Lazy Bones".

Early 1950s: Radio controlled models with single channel self built equipment became popular. In later electronic designs radios sent proportionally coded signal streams to be interpreted by a servomechanism. A servo mechanism is an automatic error sensing device used to correct the performance of a mechanism.

> Tesla is also responsible for the induction motor and the radio. Tesla's original 1897 radio patent was overturned when Marconi filed. When Marconi was awarded the Nobel Prize in 1911, Tesla unsuccessfully sued him. It wasn't until after Tesla's death that the U.S. Government reinstated the priority of Tesla's patents over Marconi probably to avoid the lawsuit by The Marconi Company against the U.S. Government for use of its patents during World War I.

1955: Zenith created the "Flash-matic" the first truly wireless remote TV controller. Since it worked by shining a beam of light onto a photoelectric cell, on sunny days, if the sun hit the TV set, it could change the channel. Maybe that was a hint that people should be outdoors enjoying the sunshine.

1956: Robert Adler developed a mechanical remote that used ultrasound to change channels and volume. When a button on the remote control was pushed it clicked. When it clicked it struck a bar. Each bar emitted a different frequency that the television detected. Adler named it the "Zenith Space Command". However, it is better known as the "clicker" because of the clicking noise it made when each button was pushed. As you may suspect, intermittent noises could change the channels.

Late 1970s: Remote television control of a much larger number of functions became available largely due to the efforts of BBC engineers and ITT using infrared communication

2002: Nintendo released the WaveBird which was the first official wireless controller made by a first party manufacturer for a video game.

Background Information

It is not surprising that remote controlled toys became very popular. They were fun to operate. The first remote controllers used actual physical wires to connect the controller to the device being controlled. The first wireless remote controllers used radio waves.

Radio control is always wireless and uses radio waves to send the signals. Radio controlled toys are labeled with the frequencies of their operation. It used to be fairly common for devices to operate on the same frequency. A duplicate controller could mess up the operation of another device. Toy manufacturers originally supplied two different versions with different frequency ranges (e.g. 27 MHz and 49 MHz) so that two models could be operated in the same area without interfering with each other.

The early garage door openers also operated on radio waves. The original wireless TV remote controllers also operated on radio waves. There were cases where a neighbor's TV remote opened and closed his neighbor's garage door. While today this sounds like a comedy routine, at one time it did happen.

Remote controls for electronic devices evolved to small hand held units with an array of buttons used to control TV, DVD players, home theatre networks and other electronic devices. The remote controllers can power on, modify the operation (change channels on TVs, pause DVD players, etc.) and power them off again all with a push of a button or two. Most of these remote controllers communicate to their respective electronic devices using infrared (IR) signals.

Infrared remote controllers can only operate over shorter distances. The typical infrared LED is low-powered and small. The typical infrared receiver on the electronics devices is small also. Radio control operates over a greater distance. The distance depends upon the power of the transmitters and receivers (antennas). Also, radio signals can go through most walls.

Both radio control signals and infrared signals travel using electromagnetic waves. They travel on different frequencies. Infrared signals have a higher frequency than radio signals.

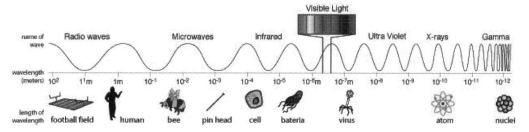

The electromagnetic spectrum from Radio waves to Gamma rays.

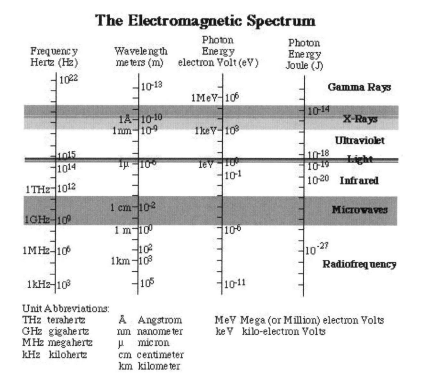

What is the Purpose of a Wireless Remote Controller?

The purpose of a wireless remote controller is to control another electrical device from a distance. The distance is totally dependent on the strength of the signal. The purpose of the electrical device being controlled depends upon what the electrical device is physically capable of doing and what it is programmed to do and what the operator of the remote controller does with the device.

Wireless remote controllers control a wide variety of devices. Wireless remote controllers are very popular in controlling electronic entertainment equipment as well as toys. There are numerous military and industrial applications as well. Wireless remote controllers are like an extra pair of extended hands.

Photo of a group of wireless remote controllers

How does a wireless radio controller work?

Let's use the example of a wireless radio controlled car whose operating frequency is 27.9 MHz and is programmed to move forward, reverse, right or left when the proper signal is received. All radio controlled toys have four parts (besides the toy itself):

- A battery (power source)

- Motor that operates the toy

- An antenna (receiver) and circuit board inside the toy that receives signals from the transmitter and activates the toy

- Hand held transmitter (controller) containing circuitry that sends radio waves to the receiver based on which button is pushed

In this example, the receiver in the radio controlled car must be turned on so that it can receive the signal from the remote controller.

There are a number of buttons on the wireless controller. Each button sends a different command to the radio controlled car. One command (button) will send the car forward. Another command (second button) will put the car in reverse. A third command (third button) will make the car turn right. A fourth command (fourth button) will make the car turn left.

A button to send a command to the radio controlled car is pressed on the radio controller. This button closes a contact in the controller completing a specific electric circuit in the controller itself. There is a different circuit for each of the four commands. The completed circuit is programmed to transmit a set sequence of electrical pulses in binary code on the operating frequency of the receiver located in the car. The transmitter sends bursts of radio waves that oscillate 27,900,000 cycles per second (27.9 MHz).

Each set sequence of electrical pulses starts with a short group of synchronization pulses to identify to the radio controlled car that a command is being sent. The synchronization pulses are the same for each button since its purpose is to notify the car that a command is coming next. The receiver on the radio car recognizes this segment. If it matches with its code, then it will accept the next group of pulses. The next group of pulses tells the receiver what action is being requested. This pulse segment is different for each button pressed.

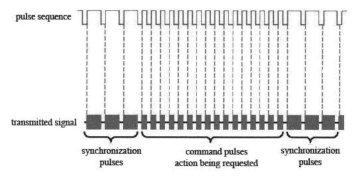

Sample of synchronized pulses sent over radio waves

If voice commands were being used, a person would say "forward", if the command is to go forward. The transmitter performs the same function by sending a group of pulses that means "forward". To make the car go in reverse, a person would say "reverse". The transmitter would send a different set of pulses that means "reverse."

The duration of the pulse tells the receiver whether the signal is a set of synchronization pulses or a set of command pulses.

Inside the car is a circuit board that controls the car. The radio receiver in the car consists of a crystal that oscillates at the frequency of the controller. When the radio receiver receives a set of electrical pulses and recognizes the synchronization segment, the appropriate circuit will activate and perform the action programmed for that particular set of electrical pulses.

Both the remote controller and the radio controlled car require energy sources. Batteries are required in both the remote controller and the car.

How does a wireless infrared controller work?

Infrared wireless remote controllers send signals using infrared radiation. Infrared radiation is red light that is in the light band that is invisible to the human eye.

Let's use the example of a TV with a wireless remote infrared controller. At the top of the remote control unit, there is a small plastic light emitting diode (LED) where the infrared radiation is emitted. On the front of the TV there's a very small infrared light detector. When the remote control is pressed, a beam of infrared radiation travels from the remote to the TV at the speed of light and the detector in the TV picks it up.

The infrared controller uses binary code to transmit commands to the TV. Binary code consists simply of "1"s and "0"s. The different combinations of "1's and "0"s are converted into numbers and letters.

Binary code	Value of binary code as a number
000	0
001	1
010	2
011	3
100	4
101	5
110	6
111	7

When a button on the remote controller is pressed, a circuit inside the controller is closed and the remote controller generates a systematic series of on/off infrared pulses in binary code. For example, a short pulse of infrared could signal a "1" and no pulse could signal a "0". Each button would close a different circuit. A sequence of infrared pulses coded for different commands is interpreted by the infrared receiver in the electronic device. Different manufacturers have set different codes to perform specific actions. Included in the sequence of pulses is a short code that identifies the device being controlled by the remote controller. That is why different controllers are needed for different devices.

A universal remote sends out signals that make almost any make or model work. A universal remote controller has to be programmed with all of the codes of all of the devices that it controls.

Applications of wireless remote controllers

The first application that comes to your mind is probably the remote TV controller or other electronic device. Next may be the remotely operated toys ranging from cars to miniature airplanes. A very common application of a wireless remote radio controller is the garage door opener.

Wireless devices that turn overhead lights on/off would save on wiring installations. There would be no need to run wires through the walls to the switches, just the wire to the device itself.

Military applications use remote devices to send out unmanned planes or boats for target practice. Radio controlled special vehicles are used for disarming bombs.

Remote operation of industrial equipment means that you can send robotic equipment into areas where it would be unsafe to send a person. Remote radio control is used in industry for overhead cranes.

How about industrial demolition? Remote devices are used to trigger explosives to implode buildings that are being demolished to make room for new buildings. It's a lot safer for the engineers to retreat a safe distance and push a wireless remote control button, than to light a fuse and hope it makes its way to the explosives.

A Japanese researcher, Kazuhiro Taniguchi, has developed a system that will enable remote operations of an IPOD by winking. The system is a single-chip computer with infrared sensors that monitors movements of the eyes. It ignores regular blinking but responds to specific eye closures. Closing both eyes for one second will start or stop an IPOD. Winking the right eye skips it forward and left eye makes it goes back.

The practical applications are endless. The hands free operation of an electronic device is like having an additional pair of hands. A hands free device can also replace the use of hands when one is temporarily or permanently disabled. This type of device could be interfaced with a number of household appliances. It would be a wonderful gift of freedom for someone like Christopher Reeve.

Learning

A wireless learning remote controller can receive and store codes that are transmitted by another remote controller. So if you purchase a new wireless remote controller and it has the learning capability, it can obtain the codes directly without you having to manually key in all of the commands. One example of this is the garage door opener that comes with your new automobile. You can download the codes directly to the new remote controller from your old controller.

Summary

Two types of wireless remote controllers are very common. In each type of device, a hand held remote controller has buttons which correspond to circuits within the device. When a button is pushed, the associated circuit closes and an electromagnet signal is transmitted. The signal contains a series of binary pulses which are in fact a coded message. The coded message contains an identifier for the device that is to receive the message. It also contains a command message for the device. The device has a receiver that is tied into electrical circuits. When the receiver receives the message, it decodes the message. If it passes the identification test, then it decodes the command message. The electrical circuit in the device is preprogrammed to take a specific action for each command that it receives. Once it decodes a legitimate command, it will perform that specific action.

The difference between the two types of wireless remote controllers is the frequency of the electromagnetic waves transmitted. A wireless remote radio controller (RC) transmits radio waves that correspond to the binary code for the button being pushed.

A wireless remote infrared (IR) controller uses the frequency of light to carry binary code for the button being pushed. The light signals are transmitted by turning the light on and off in code between the remote controller and the device that it is controlling. Infrared light is invisible to human eyes and is in the invisible portion of the electromagnetic spectrum.

In the case of the RC controller, the RC receiver decodes the radio waves and converts them into commands. In the case of the IR controller, the IR receiver decodes the light signals and converts them into commands.

The greatest advantage of radio signals is their distance range and the fact that the signals can go through most walls. That is why a lot of IF transmitters now use an RF to IR converter to extend the range of the infrared remote controller.

Concept Reinforcement

1. What is the purpose of a wireless remote controller?

2. What are the two types of wireless remote controllers?

3. What is the greatest advantage one type of wireless remote controller has over the other type?

4. Name an industrial application of the wireless remote controller.

Section 3.10 – How Things Work – Safety Sensors (Carbon Monoxide Sensors)

Section Objective

- Explain how safety sensors work

A Little History

The first carbon monoxide detector was invented in the 1960s by Figaro Engineering, a Japanese company. It used a metal oxide semiconductor (MOS) sensor.

Background Information

A sensor is any device that receives a signal and responds with a specific action based on the signal it receives.

There are a large number of sensors used today. They are everywhere…in homes, cars, offices, businesses, garages, restaurants, manufacturing facilities, airports, airplanes, shopping malls, digital cameras.

For this lesson, we will be focusing on safety sensors (specifically carbon monoxide sensors), why they are important and how they work.

Safety Sensors

The following are some of the ways that safety sensors can save lives and prevent injury:

- Carbon monoxide sensors (monitors) can save lives by alerting people to the presence of a lethal gas

- Smoke detectors can save lives by alerting people to the likelihood of a fire.

- Sensors can prevent body injuries:

  - Safety sensors should prevent elevator doors from closing when a person is standing in the opening.

  - Safety sensors should prevent garage doors from closing if there is an obstruction in the way.

  - Safety sensors should prevent automatic doors from closing if someone is standing in the open doorway

Note: one should still be careful standing in automatic doorways or elevators. Sometimes sensors malfunction and doors can close on fingers, elbows, feet or small children.

Carbon Monoxide (CO) Sensors (Monitors)

A carbon monoxide (CO) sensor is an electronic device located indoors. The device senses the presence of carbon monoxide. When the level of carbon monoxide is present in sufficient quantities to be dangerous, the device sounds an alarm.

Why is carbon monoxide called a **silent killer**? Carbon monoxide is colorless, odorless and in sufficient quantities, a lethal gas. Even low levels of carbon monoxide are deadly given sufficient length of exposure.

How Carbon Monoxide (CO) Kills

When air is breathed into the lungs, the air contains oxygen. Oxygen is absorbed into the blood and circulated throughout the body. The body removes oxygen from the bloodstream and replaces it with carbon dioxide. Carbon dioxide is breathed out from the body. The next breath brings in more oxygen. If the air contains carbon monoxide, carbon monoxide replaces the oxygen in the blood stream. If blood has a choice between carbon monoxide and oxygen, unfortunately the carbon monoxide will win. Carbon monoxide has a stronger affinity to blood than oxygen. If there is sufficient carbon monoxide in the air, the body will not be able to get enough oxygen.

The following table is a composite of information regarding exposure to toxic levels of carbon monoxide. All sources agree that the longer the exposure even to lower levels of carbon monoxide, the greater the danger of illness, permanent damage or death. Even at low levels, carbon monoxide rapidly accumulates in the body. Once carbon monoxide enters the physical body and replaces oxygen, it takes **several hours** for the blood system to rid itself of the carbon monoxide.

Carbon Monoxide Concentrations in parts per million (PPM)	Symptoms
0 to 9	Normal acceptable levels. No actions are necessary.
10 to 35	Marginal level, which could be a hazard to people with breathing problems, the elderly or infants. Action should be taken to eliminate the source of carbon monoxide and increase venting.
36 to 100	This is unhealthy and should raise a medical alert. People and pets should be evacuated and action should be taken to eliminate the carbon monoxide hazard.
100 to 199	This is dangerous. Flu-like symptoms will begin to occur. Evacuation should be immediate followed by medical attention.
200	Headache, fatigue, dizziness and nausea will increase after two to three hours of exposure at this level. Evacuation is necessary. Medical attention should be required.
800	Dizziness, nausea and convulsions will occur within 45 minutes of exposure and death will probably occur within two to three hours at this level. Immediate evacuation and medical attention are critical.
1600	Death will probably occur within an hour of exposure at this level. Immediate evacuation and medical attention are mandatory.
13,000	Death will occur within one to three minutes of exposure at this level. Anyone caught in this will need breathing apparatus to survive.

Sources of Toxic Carbon Monoxide (CO)

Carbon monoxide is the product of incomplete burning of fuels such as coal, wood, charcoal, oil, kerosene, propane or natural gas. Improperly vented or malfunctioning gas furnaces, water heaters, stoves, space heaters or wood stoves can generate carbon monoxide. Automobiles, portable generators, gas-powered tools also generate carbon monoxide. If any device or motor that gives off carbon monoxide is left running it can generate carbon monoxide fumes that accumulate. Even if a car is outside on a carport, fumes can enter a home or business through a window or vent.

Why Carbon Monoxide (CO) Sensors (Monitors) are Important

Because carbon monoxide is colorless and odorless, there is no way to recognize any level of carbon monoxide. You can't smell it and you can't see it.

Once inhaled, carbon monoxide replaces the oxygen in blood. The lack of oxygen can cause varying amounts of damage depending on the level of carbon monoxide in the air and the length of exposure. First stages are flu-like systems including headaches, shortness of breath and fatigue. Higher levels intensify the symptoms and may cause dizziness, mental confusion, even fainting. Prolonged exposure leads to brain damage and ultimately death.

The coal miners used to take a canary down into the mines. The canary with its smaller body was more susceptible to carbon monoxide poisoning. The miner would carry the canary down into the mine in a small wooden or metal cage. If the bird showed signs of distress, it was a warning that the underground conditions were not safe and the miners should immediately evacuate the mineshafts.

Carbon monoxide is known as the silent killer because its presence is so difficult to detect. Carbon monoxide sensors have saved lives and if used properly will continue to do so.

How Carbon Monoxide (CO) Sensors (Monitors) Work

Most current carbon monoxide sensors contain the following:

- A sensor capable of measuring the concentration of carbon monoxide and sending a signal when the concentration of the gas reaches a specific level.

- A microprocessor on a circuit board that receives electrical signals from the sensor and sounds the alarm.

- An alarm circuit containing a horn that is triggered by the microprocessor and is loud enough to wake people who are sleeping.

- An energy source (either a battery source, an AC plug or both (battery backup in the event of a power failure)

- Plastic container to house all of the components. This plastic container can be ceiling mounted or located elsewhere for optimal safety performance.

- Optional: a visual display that shows the actual level of carbon monoxide at that specific location.

- Optional: a horn alarm silence feature. This is to shut off the sound but it doesn't mean the alarm should be ignored. If the horn goes off, the area should be evacuated immediately.

Photo of a carbon monoxide sensor

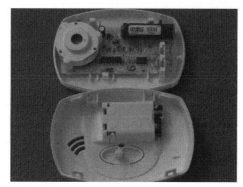

The inside of a carbon monoxide sensor

The type of sensor in a carbon monoxide sensor is the most important design factor.

The simplest type is known as a detection card. The card is printed with a dot that changes color based on a chemical reaction when it comes into contact with carbon monoxide. Depending on the design of the sensor, this type of detector may not sound an alarm and requires regular checks to determine if it has been exposed to carbon monoxide. An audible alarm is generally a safer feature. However, for the hearing impaired, the visual display is better than nothing.

There is a bio-mimetic gel sensor named so because it is designed to mimic the body's response to carbon monoxide by continually absorbing gas. But because it cannot reset itself when the carbon monoxide is dispelled, it continues to build up carbon monoxide and give off false alarms. False alarms over time tend to be ignored (not a good thing). Resetting the bio-mimetic gel sensor after an actual exposure can take up to 48 hours. During that time, the sensor is not functioning properly and another actual exposure may not be recognized by the sensor.

Metal oxide sensors are more accurate and commonplace. The metal oxide sensors use solid state tin-dioxide circuits, which continually monitor the air for the presence of carbon monoxide. These sensors can have a digital readout display showing the actual carbon monoxide levels in real time. When the carbon monoxide level reaches a predetermined level, an audible alarm is sounded.

Carbon monoxide sensor with digital display

When a specific carbon monoxide level is reached, the detector sounds an alarm. Unfortunately, the detector may be sensitive to gases other than carbon monoxide that are found in the home such as hair spray. The accuracy of this type of sensor can drift significantly after being in use for a while.

More effective and long lasting is the Instant Detection and Response (IDR) electrochemical sensing technology. IDR technology is used as an industry standard for professional sensing equipment and will instantly detect the presence of carbon monoxide. Detectors built with this technology will not react to other gases and are accurate to within plus or minus 3%.

The energy source for these detectors is very important. Both battery-operated and alternating power (AC) detectors are available. Like any device that uses batteries, the batteries must be replaced periodically or the sensors will not work.

Some devices can be hard wired. Multiple detectors can be hard wired together into the incoming electrical circuit. If any one of the sensors detects carbon monoxide, they can all sound the alarm. Not a bad idea if there are suspected or known carbon monoxide concerns throughout an area. This could be important in some manufacturing facilities.

Industrial carbon monoxide sensors (monitors) may have a flashing light tied into the audible alarm. Neither the light nor the alarm should ever be disabled. The light helps notify personnel if the environment is noisy and earplugs are in use. However, the audible alarm should be loud enough to be heard even with ear plugs.

For home use, some units come with both AC powered plug-ins and battery backup. If there is a power failure, the battery can back up the unit and still protect the area. Since people tend to use alternate means of heat and power when electricity goes out, and these devices could be potential carbon monoxide sources, having battery backup is a really good idea. That is, as long as the batteries are installed and working!

Where Carbon Monoxide Sensors should be installed

Carbon monoxide's specific gravity is 0.9657 and a normal mixture of air's specific gravity is 1.00. So carbon monoxide is roughly the same weight as air.

Carbon monoxide may be contained in warm air coming from combustion appliances such as a furnace. Carbon monoxide will rise with the warmer air. That is why carbon monoxide sensors should not be installed directly above any fuel-burning appliance or within fifteen feet of heating or cooking appliances. Small amounts of carbon monoxide may be emitted upon start-up of these appliances and provide unnecessary alarms. False alarms are a nuisance. Like the young child who called "wolf" too many times in order to gain attention, when the real wolf showed up, he was ignored. People tend to ignore real alarms if they are mixed in with too many false alarms. In this case, like in the case of the wolf, it could be fatal.

Humid places such as bathrooms are also not a good place to install carbon monoxide sensors. Kitchens and garages are also not recommended for carbon monoxide installations although both are good locations for smoke detectors.

For a carbon monoxide sensor to provide optimum coverage, the proper placement of the sensor is important. The best place in a home to install a carbon monoxide sensor is in every bedroom where people are sleeping. If the home is multi-level, additional monitors on each level are recommended. Mounting the carbon monoxide sensors on the ceilings is recommended by most manufacturers. While carbon monoxide monitors should not be mounted directly above fuel-burning appliances, there should be one at least 20 feet (about 6 meters) away from them.

Applications of Safety Sensors

Today

We've discussed in detail carbon monoxide detectors. Smoke detectors are also very important to safety. Smoke detectors are similar in functionality and should be located in each bedroom and on each level of a house.

Other types of sensors include optical sensors. Some manufacturing facilities use automatic guided vehicles (AVGs) to move materials. Located indoors in a warehouse using preprogrammed paths, the AVGs can be programmed to follow set routes. Photoelectric sensors mounted on the front and back of the AGVs are used to help prevent the AVG from colliding with objects or persons in their path.

Sensors make things more convenient for people. An example is the automated door opener. They also keep the door from closing on a person.

Future

You've seen science fiction movies featuring high speed cars that run on autopilot. The cars travel at incredible speeds and never get into an accident. The cars sense oncoming obstacles that swerve, slow down or stop to avoid collisions. Wouldn't it be great if all cars were equipped with sensors that prevented accidents? No one would be hurt or killed by car crashes. Well, the technology is here. It is only a matter of time and money before cars can have safety sensors installed that will signal the driver of an oncoming obstacle. Or better yet, have the sensor stop the vehicle before it hits anything.

Note: One of the car manufacturers now advertises an automatic stop if the car senses something behind the car when the car is moving in reverse.

As gas sensing electronics improve, future detectors could be tied into household appliances. If a household appliance such as a gas furnace or gas hot water heater generated too much carbon monoxide, the sensor can send a signal to shut down the flow of gas to the appliance so no further carbon monoxide is released. This potentially could be a life saving feature. If however, the sensor is malfunctioning, this could be a real nuisance. It is important to insure that the technology works properly. Otherwise, the tendency is to shut off the safety feature rather than solve the problem.

Summary

Safety sensors are devices that receive a signal and provide a preprogrammed response. They can be complicated or simple depending upon the application. Safety sensors are designed and implemented in order to reduce illness, injuries and death. Bypassing safety sensors is never a good idea. If a safety sensor triggers an alarm, the most important thing is to identify the alarm condition and take the appropriate action. If it is a carbon monoxide warning, immediate evacuation is critical. If it is a smoke alarm, again immediate evacuation is critical. Then call for help.

It is probably a good idea to implement the simplest solutions to monitor unsafe conditions. Sound audible alarms, light lights, ring bells or whatever it takes to get someone's attention. Then rely on human action to clear out an area. Only expert help with appropriate equipment should go back and fix the problem.

Concept Reinforcement

1. What is the definition of a sensor?

2. True or False. Carbon monoxide is a colorless, odorless, potentially lethal gas.

3. True or False. Continued exposure to carbon monoxide is accumulative (the amount of CO continues to increase in the blood over time)

4. True or False. Carbon monoxide at the level of 1600 PPM (parts per million) is not lethal.

Section 3.11 – How Things Work – Direct Current Motors

Section Objective

- Explain how direct current motors work

Purpose

This section will focus on motors that operate using direct current (DC). The purpose of the section is to learn about direct current motors. You should understand the basic components of direct current motors and how they operate. You should also be familiar with the three main types of direct current motors. You should be able to calculate the horsepower output of a motor and determine its efficiency.

A Little History

1821: British scientist Michael Faraday demonstrated the principle of converting electrical energy into mechanical energy. Faraday placed a permanent magnet into a pool of mercury. Then he lowered a free-hanging wire until the end was in the mercury. Faraday passed an electric current through the wire. The wire rotated around the magnet showing a circular magnetic field existed around the wire.

1825: British scientist William Sturgeon invented the electromagnet. He wrapped a loosely wound coil around a horseshoe shaped piece of iron. When he passed an electric current through the wire, it became a magnet. When he stopped the current, it no longer exhibited the properties of a magnet.

1828: Hungarian Anyos Jedlik demonstrated the first real electric motor using electromagnets.

1832: British scientist William Sturgeon invented the commutator and built a workable direct current electric motor.

1837: American Thomas Davenport patented a direct current electric motor using a commutator intended for commercial use. Davenport built several motors, but due to the high cost of using battery power for the energy source, the motors were not economically feasible.

1869: Belgium electrical engineer Zénobe Gramme invented the Gramme Dynamo which is a continuous-current electrical generator. The Gramme Dynamo produced higher voltages than earlier dynamos.

1873: Zénobe Gramme accidentally found that the generator was reversible and could be used as an electric motor. This was the first electrical motor with sufficient power to be useful in industrial applications.

A Quick Review

- **Current** is the flow of electrons along a path. An **electrical circuit** is a complete path along which electrons flow with a return to the source.

- **Electromagnetic force (EMF)** is the electrical pressure that pushed electrons through a conductor.

- By convention, the **direction of the magnetic field** is from N to S.

- By convention, the **direction of the flow of electric current** goes from plus (+) to minus (-).

- **Negatively charged electrons move from (-) to (+).** Electrons move in the opposite direction than the convention for the direction of current.

- The **ampere** (amp) is a measurement of the amount of electricity that flows through a circuit. An ampere or amp is represented by "I" or "A".

- **Voltage** is the electrical pressure which pushes electrons through an electric circuit. Voltage is the difference in electrical potential between two points in a circuit and is measured in volts. Volts are represented by "V" or "E."

- **Voltage drop** is the difference in voltage from one end of an electrical circuit to the other.

- **Wattage** is the amount of power that is used in an electric circuit. The unit of measurement is the watt and is represented by "W" or "P." One watt is one ampere of current flowing at one volt.

- **Power** is the rate at which work is performed. Electric power is the rate electrical energy is transferred by an electric circuit. Units of measurement are watts "W" and horsepower "hp." (1 horsepower = 746 watts) (1 watt = 0.0134 horsepower)

- **Work** is power used over a period of time. Units of measurement are joules, BTUs and calories. (1 BTU = 1050 joules) (1 calorie = 4.186 joules)

- Work and energy are basically the same and use the same units of measurement (joules, BTUs, and calories).

Definitions used in the Explanation of Direct Current Motors

- An **armature** is the rotating part of a motor, typically made of copper wire wound around an iron core.

- **Armature reaction** is the bending or twisting of the main magnetic field of the motor.

- A voltage is induced into the armature of a direct current motor when the windings of the armature spin through the magnetic field produced by the pole pieces. This induced voltage is opposite in polarity to the applied voltage and is called **back-EMF (electromagnetic force) or counter-EMF.**

- A **commutator** is an electrical switch that periodically reverses the current flowing within an electric motor. In combination with brushes (commonly made of carbon), the commutator provides the electrical connection between the pole pieces (stationary terminals) and the rotating armature.

- **Field windings** are wound electromagnets that produce the magnetic field when current is applied. Shunt field windings are connected in parallel with the armature. Series field windings are connected in series with the armature.

- An **Interpole** is a small pole piece placed between the main field poles in a direct current motor. The purpose is to reduce armature reaction. They are usually made of the same material as the main pole pieces.

- **Newton** is a unit of force (named after Sir Isaac Newton in recognition of his work in mechanics). The unit is represented by "N." (1 N = 0.22481 pound-force)

- Main **Pole pieces** (stationary terminals) provide the magnetic field in a direct current motor. Pole pieces are made of some type of good magnetic conductive material and are located inside the housing of the direct current motor.

- **Rectification** is the process of converting alternating current (AC) to direct current (DC). The device that performs the conversion is called a **rectifier**. This device would permit the operation of a direct current motor using power supplied from a power grid.

- A **rheostat** is a variable resistor.

- **RPM** is the number of revolutions per minute that the shaft on a motor turns.

- **Shaft load** is the amount of resistance to the shaft turning when the direct current motor starts running or is running. It is the load (physical device) attached to the shaft. When the motor runs, the shaft turns, and the device powered by the motor performs its function.

- **Torque** is a turning force. In this application, torque is the rotational force down the shaft. The force of the turn determines how big a load the shaft will turn. Units of measurement are foot-pound "ft-lb" and newton-meter "Nm." (One newton is the amount of force that, when applied to 1 kg of mass, will give it an acceleration of 1 m/sec^2)

What is a Direct Current Motor?

A direct current motor is a rotating device energized by direct current (DC) used to convert electric energy to mechanical energy. There are three basic types of direct current motors:

- shunt direct current motor–the field winding is placed in parallel with the armature

- series direct current motor–series field winding is connected in series with the armature

- compound direct current motor–series field winding connected in series with the armature and shunt field winding in parallel with the armature (combination of shunt and series)

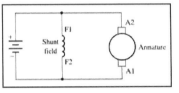

Shunt Motor Electrical Diagram

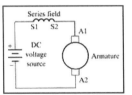

Series Motor Electrical Diagram

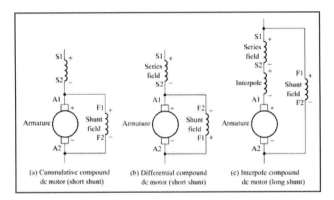

(a) Cummulative compound dc motor (short shunt) (b) Differential compound dc motor (short shunt) (c) Interpole compound dc motor (long shunt)

Electrical diagram of a shunt motor, a series motor and compound motors

What's in a Direct Current Motor?

Direct current motors contain the following:

- Armature – rotor-mounted windings (rotates inside magnetic field)

- Field poles – stationary windings

- Interpoles – additional field poles (only on compound interpole motors)

- Commutator – an electrical switch that reverses flow of current and connects the field poles to the armature in combination with brushes

- Brushes – provide power connection from outside power source to the armature

- Shaft – drives the load

- Electrical plug to connect to a direct current power source or wires to connect directly to a battery.

- On/Off switch to turn the motor on and off

- Housing to contain the motor

How a Direct Current Motor Works

The purpose of a direct current motor is to turn a shaft. The shaft is attached to a load. For example, the load might be a drill. When the motor is turned on, it draws current. The armature turns and turns the shaft. The shaft turns and turns the drill bit.

The direct current motor consists of rotor-mounted windings (armature) and stationary windings (field poles). The commutator is attached to the armature which is mounted on the rotor. Carbon brushes and the commutator function as a sliding switch.

As the commutator rotates, the carbon brushes slide over the commutator's copper surfaces. When current is supplied from the direct current power supply, the current is conducted through the brushes and commutator to the armature windings.

The combination "switch" energizes portions of the armature based on the position of the rotor (the rotating windings that make up the armature).

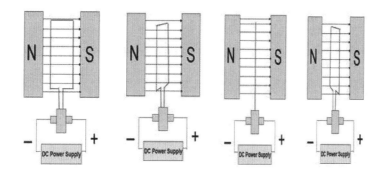

Simplistic diagram of a direct current motor

Current is passed through the armature field windings (wires) becoming electromagnets. As the armature (which is carrying current) passes between the stationary poles, a magnetic field is created between the two stationary poles.

The magnetic attraction and repulsion causes the armature (rotor) to rotate. The outer magnets repel the electromagnets and rotate the shaft. The armature rotates perpendicular to a constant magnetic field. The direction of the magnetic field is always in one direction (north to south). Once the armature makes a 180 degree turn, it will be in the opposite direction. For the armature to continue to turn in the same direction, the current flowing through the wire must be reversed every half turn.

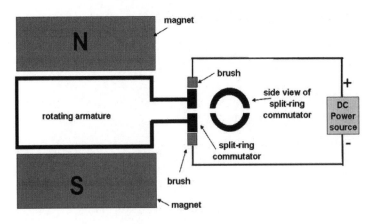

Drawing of a split-ring commutator

The split-ring commutator reverses the polarity of the armature (coil) with respect to the external circuit when the armature passes through the plane perpendicular to the direction of the magnetic field. Split-ring commutators are necessary for a direct current motor.

As soon as the armature starts rotating, it produces a voltage that is opposite in polarity to the power supply. This voltage is called back-EMF or counter-EMF. The back-EMF voltage reduces the overall voltage so that the armature "sees" a smaller voltage potential.

The Shunt Direct Current Motor

The direct current shunt motor has the shunt field connected in parallel with the armature. The shunt winding consists of many turns on the coil with small-gauge wire. Because the wire is small and many more turns of the wire are feasible, it can produce a very strong magnetic field but it cannot produce a large current. This means the shunt motor has a low starting torque which will only handle a small shaft load.

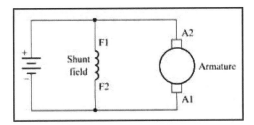

Electrical diagram of a shunt motor

When voltage is applied, the high resistance of the shunt coil supplies a low current flow. The armature will draw current and produce a magnetic field strong enough to cause the armature shaft with its load to start rotating. When the armature begins to turn, it will produce back EMF.

The back EMF will reduce the current in the armature. The amount of current the armature will draw is directly related to the size of the load when the motor reaches full speed. When the motor reaches its maximum speed (RPM), the speed will remain fairly constant.

Increasing the load slightly while at full RPM

As soon as the armature starts turning, it starts to produce back-EMF. After the motor reaches its desired speed, increasing the load slightly causes the armature shaft to slow down and less back-EMF will be produced. The difference between the back-EMF and the voltage applied to the circuit will become larger (applied voltage remains the same). The effective voltage is increased. This will cause more current to flow. The extra current produces an increase in the magnetic field strength. The result is the armature provides the motor with the extra torque needed to regain its RPM when the load is increased slightly.

The shunt motor will run at less than full RPM when the voltage supplied is less than its rated voltage. At the reduced voltage, the motor will have a tendency to overheat and operate at a lower torque. It will lose efficiency under these operating conditions. Shunt motors are designed to run most efficiently at their full rated RPM.

The Series Direct Current Motor

The advantage of the series motor is that is has the capability to move very large shaft loads when it is first energized. The series motor provides high starting torque.

In the series motor, the field winding is connected in series with the armature. Because it will carry the same current that passes through the armature, the field wiring is made from heavy-gauge wire that is large enough to carry the load. Because the wire is so large, the winding will have few turns of the wire.

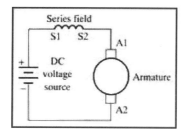

Electrical diagram of a series motor

The amount of torque the motor can produce is dependent upon the amount of current that passes through the field windings. Large conductors carry large amounts of current which results in the production of large torques.

Series motors are best used to start car engines and other applications where the series motor does not operate for an extended period of time. They do draw large amounts of current when running.

When the voltage is first applied to the circuit, the armature is not rotating. The only resistance in this circuit will be provided by the large conductors in the armature and field windings. Because the conductors are large, the resistance will be small. The motor will draw a large amount of current. When the current begins to flow through the field windings and armature, a strong magnetic field will occur.

The strong magnetic field will cause the armature to spin with the maximum amount of power creating a large torque. As the armature spins, it begins to produce voltage which is opposite in polarity to the power source. This is back-EMF (electromotive force). The amount of back-EMF will effectively reduce the supply voltage to the motor windings.

Decreasing the voltage in a series motor

According to Ohm's law, a slightly lower voltage means a slightly reduced current. Reduced current means the motor will lose torque as the motor speed increases. However, the load is already moving when the armature begins to pick up speed so less torque is required to keep the load moving. The motor continues to operate its load but with less current. This helps prevent overheating.

Losing the load at full speed of series motor

If the series motor is running with full load and it accidentally loses its load, the current falls to a minimum. Since the armature would not produce enough back-EMF, the amount of back-EMF would be reduced. With the load not present on the shaft, the armature would increase its rotational speed. It could become a runaway motor unless some type of runaway protection was installed.

Controlling speed is difficult in a series motor

It is difficult to control the speed of a series motor by external means because the RPM of a series direct current motor is determined by the size of its load.

The Compound Direct Current Motor

The compound motor is a combination of the shunt motor and the series motor. A shunt field winding is connected in parallel with the armature and a series field winging is connected in series with the armature. Because both shunt and series windings are part of the combination motor, the best characteristics of the shunt and series motors are included in the combination motor. The compound motor has the regulated speed characteristics of the shunt motor and the torque capabilities characteristic of the series motor.

There are three basic types of compound motors:

- Cumulative compound direct current motor
- Differential compound direct current motor
- Compound interpole direct current motor

Cumulative Compound Direct Current Motor

The cumulative compound motor is connected so its coils aid the magnetic fields of both the series field and armature. The shunt winding can be wired as a long shunt or as a short shunt.

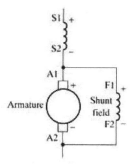

Electrical diagram of a cumulative compound motor short shunt

Differential Compound Direct Current Motor

Differential compound motors use the same motor windings as the cumulative compound motor, but they are not connected the same way. The differential compound motor is connected so its polarity is reversed to the polarity of the armature. When the shunt field's polarity is reversed, its field will oppose the other fields.

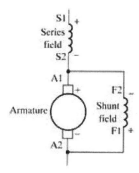

Electrical diagram of a differential compound motor

The differential compound motor will not have the speed control that is characteristic of a shunt motor by itself. When the differential compound motor is running at full speed and the load on the motor is reduced, the differential compound motor will tend to runaway like the series motor. When the differential compound motor is running at full speed and the load is increased, the speed will drop more than the cumulative compound motor.

Compound Interpole Direct Current Motor

The compound interpole motor has interpoles connected in series with the armature and series windings. An interpole is made out of the same gauge wire as the series winding. It is connected so that its polarity is the same as the series winding pole it is mounted behind. Depending on the motor design, the motor may have a number of interpoles to make the field stronger.

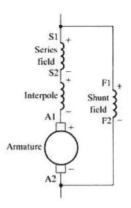

Electrical diagram of a compound interpole motor

The armature reaction is the twisting or bending of the main magnetic field of the motor. Armature reaction is increased with increased current passing through the armature. When current is flowing through the armature, the armature rotates through the magnetic field. The movement of the armature bends or twists the magnetic field. Interpoles have the function of reducing the armature reaction effect in the commutating zone by making the magnetic field stronger. Interpoles help prevent the armature and brushes from arcing due to the buildup of magnetic forces.

Since the brushes do not arc, they will last longer and not need to be replaced as frequently. The interpoles also allow the armature to draw heavier currents and carry larger shaft loads.

Calculations

How to Calculate the Horsepower (hp) required in an Application

You need to know the following before you can determine the horsepower of the motor required:

- Voltage source
- Speed (RPM) to drive load
- Torque (ft-in) required to drive the load at desired speed

hp = (1.59 × torque × RPM)/100,000

where 1.59 and 100,000 are constants

Example: Using the above formula, how much horsepower is required by a motor running at 1500 RPM and turning a load of 325 lb-in.?

hp = (1.59 × torque × RPM)/100000

hp = (1.59 × 325 × 1500)/100000

hp = 775125/100000

hp = 7.75

Note: when calculating hp, verify that the constants are applicable to your motor.

How to Calculate the Efficiency of a Running Motor

First calculate the output horsepower of a motor.

Convert the horsepower into watts for the output power.

Calculate the input power: watts = volts × amps

The efficiency of a motor can be found using the following formula:

Efficiency = (power out/power in) x 100

Example: Using the formulas given, what is the efficiency of a motor running at 1200 RPM, turning a load of 8 lb-in, using 120 volts and drawing 1.3 amps?

hp = (1.59 × torque × RPM)/100000

hp = (1.59 × 8 × 1200)/100000

hp = 15264/100000

hp = 0.153 (rounding numbers)

Convert hp to watts (1 hp = 746 watts)

0.153 hp × 746 watts/hp = 114.1 watts (rounding numbers)

Efficiency = (power out/power in) × 100

Efficiency = (114.1/ (volts x amps)) × 100

Efficiency = (114.1/ (120 x 1.3)) × 100

Efficiency = (114.1/156) × 100

Efficiency = 0.73 × 100

Efficiency = 73%

Disadvantages of Direct Current Motors

Some of the disadvantages of using direct current motors are:

- Economics plays a major factor in the selection of motors. One of the disadvantages of direct current motors is the cost of the power supply as compared to alternating current motors which can be powered from the power grid.

- Power supplied from the power grid must be rectified (converted from AC to DC) in order to operate a direct current motor.

- The conversion of alternating current (AC) to direct current (DC) has been known to cause power quality problems. Frequently direct current motors are more complicated and expensive to manufacture than the equivalent alternating current motor.

Advantages of Direct Current Motors

Some of the advantaged of using direct current motors are:

- The speed of shunt direct current motors is easy to control by adjusting the field current and/or the armature voltage.

- The series-connected motor has the ability to produce large torque under load without requiring excessive current.

- Direct current motors run on batteries.

- Compound motors combine the best of the shunt motors and series motors.

Summary

A direct current motor converts electrical energy into mechanical energy. The direct current motor contains a rotating part called an armature. The armature is made of a conductive material. When current flows through the armature, it produces a magnetic field around it. The armature is physically between field windings, which are also made of conductive material. When a current carrying wire passes between two conductors, a magnetic field is created between the conductors. When current is applied to the armature, the armature produces a magnetic field around itself and at the same time creates a magnetic field between the field windings. One side of the armature is attracted to the field windings and the other side is repulsed. This causes the armature to rotate through the magnetic field.

The armature is attached to a shaft, which turns when the armature turns (current flows through the armature). The shaft is attached to a load. **Turning (driving) the load is the purpose of the motor.** The strength of the torque (turning of the shaft by the motor), determines how much of a load the motor can drive.

Two things determine the amount of torque produced by a motor:

- The strength of the magnetic field produced by the pole pieces.
- The strength of the magnetic field of the armature.

There are three main types of direct current motors:

- **Shunt direct current motors**–the field wirings are high-resistance and are connected in parallel with the armature. Speed is easily controlled in these motors. The speed can be changed by varying the field resistance or voltage supplied.

- **Series direct current motors**–the field windings are in series with the armature. The speed of series motors is difficult to control. The series direct current motor will develop the highest starting torque. They are ideal for electrical devices that require a high starting torque such as an automobile starter.

- **Compound direct current motors**–combine the best of both shunt and series motors. The compound direct current motor will develop less torque than the series direct current motor but more than the shunt direct current motor. There are three basic types of compound direct current motors:

    Cumulative compound direct current motor–connected so its coils aid the magnetic fields of both the series field and armature.

    Differential compound direct current motor–connected so its polarity is reversed to the polarity of the armature.

    Compound interpole direct current motor–has interpoles connected in series with the armature and series windings to make the magnetic field stronger.

Efficiency of Motors

The efficiency of a motor can be calculated if you know the voltage supplied, the current drawn and the actual horsepower output of the motor. Using Ohm's Law to calculate the power in where power in = voltage × current, then efficiency can be calculated by:

Efficiency = (power out/power in) × 100

Concept Reinforcement

1. What are the three types of direct current motors discussed in this lesson?

2. What is the formula to calculate the efficiency of a motor?

3. True or False. It is difficult to control the speed of a series direct current motor.

4. A compound motor is the combination of a series direct current motor and a _____ direct current motor.

Section 3.12 – How Things Work – Alternating Current Motors

Section Objective

- Explain how alternating current motors work

Purpose

This section will focus on motors that operate using alternating current (AC). The purpose of the section is to learn about alternating current motors. You should understand the basic components of alternating current motors and how they operate. Because there are so many different types of alternating current motors, this lesson will focus on the three types of three-phase alternating current motors. You should be able to calculate the horsepower output of a motor and determine its efficiency.

A Little History

In **1888,** Nikola Tesla invented the first practicable alternating current (AC) motor and with it the polyphase power transmission system. Tesla continued his work on the alternating current (AC) motor in the years to follow at the Westinghouse Company.

A Quick Review

- **Current** is the flow of electrons along a path. An **electrical circuit** is a complete path along which electrons flow with a return to the source. In **alternating current**, the direction of the flow of electrons is reversed at regular intervals.

- **Electromagnetic force (EMF)** is the electrical pressure that pushed electrons through a conductor.

- By convention, the **direction of the magnetic field** is from N to S.

- By convention, the **direction of the flow of electric current** goes from plus (+) to minus (-).

- **Negatively charged electrons move from (-) to (+).** Electrons move in the opposite direction than the convention for the direction of current.

- The **ampere** (amp) is a measurement of the amount of electricity that flows through a circuit. An ampere or amp is represented by "I" or "A."

- **Voltage** is the electrical pressure which pushes electrons through an electric circuit. Voltage is the difference in electrical potential between two points in a circuit and is measured in volts. Volts are represented by "V" or "E."

- **Voltage drop** is the difference in voltage from one end of an electrical circuit to the other.

- **Wattage** is the amount of power that is used in an electric circuit. The unit of measurement is the watt and is represented by "W" or "P." One watt is one ampere of current flowing at one volt.

- **Power** is the rate at which work is performed. Electric power is the rate electrical energy is transferred by an electric circuit. Units of measurement are watts "W" and horsepower "hp." (1 horsepower = 746 watts) (1 watt = 0.0134 horsepower)

- **Work** is power used over a period of time. Units of measurement are joules, BTUs and calories. (1 BTU = 1050 joules) (1 calorie = 4.186 joules)

- Work and energy are basically the same and use the same units of measurement (joules, BTUs, and calories).

Definitions used in the Explanation of Alternating Current Motors

- **Field windings** are wound electromagnets that produce the magnetic field when current is applied.

- **Newton** is a unit of force (named after Sir Isaac Newton in recognition of his work in mechanics). The unit is represented by "N". (1 N = 0.22481 pound-force)

- **Pole pieces** (stationary terminals) provide the magnetic field in an alternating current motor. Pole pieces are made of some type of good magnetic conductive material and are the stators located inside the housing of the alternating current motor.

- A **rotor** is the rotating part of an alternating current motor.

- **RPM** is the number of revolutions per minute that the shaft on a motor turns.

- A **stator** is the stationary part of an alternating current motor.

- **Shaft load** is the amount of resistance to the shaft turning when the alternating current motor starts running or is running. It is the load (physical device) attached to the shaft. When the motor "runs", the shaft turns, and the device powered by the motor performs its function.

- **Synchronous speed** is the speed of the rotating magnetic field of an alternating current (AC) induction motor.

- **Torque** is a turning force. In this application, torque is the rotational force down the shaft. Units of measurement are foot-pound "ft-lb" and newton-meter "Nm". (One newton is the amount of force that, when applied to 1 kg of mass, will give it an acceleration of 1 m/sec^2)

What is an Alternating Current Motor?

There are two major categories of alternating current motors.

- Three-phase motors
- Single-phase motors

Note: Two-phase power is produced in some parts of the world. Two-phase motors will not be addressed in this section.

Note: There are many different types of single-phase motors with unique operating principles.

There are three basic types of three-phase motors.

- Squirrel-cage induction motor
- The wound rotor induction motor
- The synchronous motor

This section will focus on one of the three-phase motors: the squirrel-cage induction motor.

Three-phase alternating current motors convert three-phase alternating current into mechanical energy. Three-phase alternating current motors are usually lighter, smaller and more efficient than single-phase motors. They operate all types of machinery.

What's in an Alternating Current Motor?

Alternating current motors contain the following:

- Rotor – rotating part
- Stator – stationary winding
- Shaft – drives the load
- Electrical plug to connect to the correct alternating current power source using the required voltage (V) and frequency (Hz)
- On/Off switch to turn the motor on and off
- Housing to contain the motor

How an Alternating Current Motor Works

The purpose of an alternating current motor is to turn a shaft. The shaft is attached to a load. For example, the load might be a conveyor belt. When the current is applied to the motor and the motor is turned on, the rotor in the motor turns. The rotor is attached to the shaft. When the rotor turns, the shaft turns and the conveyor belt moves.

The stator is the stationary part of the alternating current motor. The stator contains a number of windings (poles). Inside the stator is a rotor. A rotor is the part of the alternating current motor that rotates inside the stator. Both rotors and stators are made of conductive material.

When current is applied to any conductor and that conductor passes between two additional conductors, a magnetic field is produced between the two additional conductors. There is also a magnetic field around the first conductor because it is carrying current. The two magnetic fields begin interacting with each other.

The stator contains a number of poles (field windings) of conductive material around the inside. When current is applied to the rotor (a conductor) it creates a magnetic field around itself. Because it is inside a stator which contains a number of poles of conductive material, as the current passes through the rotor, a magnetic field is set up within the inside of the stator. As a result, the magnetic field of the rotor is attracted to the magnetic field of the stator and the rotor begins to turn in the same direction as the rotating magnetic field.

The one thing that makes all of the three-phase alternating current motors work is the rotating magnetic field.

So what is a rotating magnetic field? The current is alternating current and it is also three-phase. The voltages supplied in a three-phase system are 120° out of phase with each other. The three voltages change polarity at regular intervals.

The primary force of the magnetic field rotates as the power cycles. The speed at which the magnetic field rotates is called the synchronous speed. Two things affect the synchronous speed:

- The frequency of the applied voltage (50Hz or 60Hz)
- The number of stator poles per phase

The rotor is attached to the shaft. The rotor is turned as the result of electric power. The rotor converts the electric power into mechanical power.

The major difference in the types of three-phase alternating motors is the way power is supplied to the rotor. There are three basic types of three-phase motors:

- Squirrel-cage induction motor
- Wound Rotor Induction Motor
- Synchronous Motor

Motors

Squirrel-Cage Induction Motor

The squirrel-cage induction motor gets its name from the type of rotor used in the motor. The rotor would look like a squirrel cage if the laminations were removed. It is made by connecting conductive bars to two end rings.

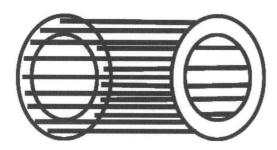

Basic squirrel-cage rotor

The squirrel-cage motor is an induction motor. In an induction motor, the current flow in the rotor is produced by induced voltage from the rotating magnetic field of the stator. When power is first connected to the stator, the rotor is not turning. The rotating magnetic field of the stator cuts the rotor bars. The cutting action is what induces the voltage into the rotor bars. Cutting action means the magnetic field "cuts" through the rotor. Current flows through the rotor bars as a result of the induced voltage. As current flows through the rotor, a magnetic field is formed around each of the rotor bars. The magnetic fields of the rotor are attracted to the magnetic field of the stator. The rotor starts turning in the same direction as the rotating magnetic field of the stator.

When the squirrel-cage motor is started, the current draw to get it started is several times greater than its normal running current.

To determine the starting current required, you would need to know the horsepower (hp) rating of the motor, the type of rotor bars, and the applied voltage. There is a code letter provided by the National Electrical Code that gives the kilovolt-amperes per horsepower required for the locked rotor current for each type of rotor bar used in industry. The locked rotor current would be the current draw when the motor is running in steady state (running normally with no increase or decrease of load or change in power supply).

Example: A 600-hp, three-phase squirrel-cage induction alternating current motor was selected. It will be connected to 2000V. The motor has a code letter of H. From the table, H has a value from 6.3 to 7.09 kilovolt-amperes/ horsepower.

Calculate the average value of H:

(6.3 + 7.09)/ 2 = 13.19/2

Average value of H = 6.695 kilovolt-amperes/ horsepower (kVA/hp)

Rounding to the nearest decimal point, H = 6.7 kVA/hp

The apparent power requirements (kilovolts-Amps) can be calculated by multiplying the factor H by the horsepower rating of the motor:

$kVA_{motor} = H \times 600 \text{ hp}$

$kVA_{motor} = 6.7 \text{ kVA/hp} \times 600 \text{ hp}$

$kVA_{motor} = 4020 \rightarrow VA_{motor} = 4,020,000$

$I_{line} = 1.732 \times$ phase current (standard number)

$I_{line} = VA/ (V \times 1.732)$

$I_{line} = 4020000/ (2000 \times 1.732)$

$I_{line} = 4020000/3464$

$I_{line} = 1160.5$ amps the motor would need this much current to start.

The amount of induced voltage is determined by the following:

- The strength of the stator's magnetic field
- The number of bars in the squirrel-cage rotor
- The speed of the cutting action (as the speed of the rotor increases, the rotating magnetic field cuts the rotor bars at a slower rate).

The amount of torque produced by the induction motor is directly related to the strength of the magnetic field of the rotor. Three factors affect the amount of torque:

- The strength of the magnetic field of the stator
- The strength of the magnetic field of the rotor
- The phase angle difference between rotor and stator fields.

However, an induction motor can never reach synchronous speed. If the rotor turns at the exact same speed as the rotating magnetic field, the rotor would not have any induced voltage. That means there would be no current in the rotor. If the rotor does not carry any current, it would not have a magnetic field. Without a magnetic field, there would be no torque. Without torque, there would be no turning of the shaft and there would be no load.

Wound Rotor Induction Alternating Current Motor

The wound rotor induction alternating current motor is similar to the squirrel-cage induction alternating current motor. The difference between the two is the construction of the rotor. Unlike the squirrel-cage rotor, the wound rotor is constructed by winding three separate coils on the rotor 120° apart.

Synchronous Alternating Current Motor

The synchronous alternating current motor is not an induction motor. It will operate at the speed of the rotating magnetic field. Its speed will be constant from load to no load. However, the synchronous alternating current motor does require help during startup from a direct current supply in order to reach the synchronous speed required.

It starts out running like the squirrel-cage induction alternating current motor. However, once the rotor has accelerated almost to its synchronous speed, direct current (DC) is applied to the rotor and the windings of the rotor become electromagnets. The electromagnetic field of the rotor then syncs with the rotating magnetic field of the stator.

Summary

There are single-phase, split-phase and three-phase alternating motors. There are induction motors and non-induction motors. Focusing on three-phase motors, there are three basic types:

- Squirrel-cage induction motor
- Wound Rotor Induction Motor
- Synchronous Motor

In each motor, there is a stationary part called a stator inside which are mounted poles. These poles are field windings of conductive material. They are spaced around the inside of the stator. Inside the stator is a rotor. A rotor is the part of the alternating current motor that rotates inside the stator. The rotor is also made of conductive material.

The one thing that makes all of the three-phase alternating current motors work is the rotating magnetic field.

The current is alternating current and it is also three-phase. The voltages supplied in a three-phase system are 120° out of phase with each other. The three voltages change polarity at regular intervals. The primary force of the magnetic field rotates as the power cycles.

The focus in this lesson was on the squirrel-cage induction motor. The squirrel-cage induction motor uses a rotor that consists of parallel bars with rings on each end.

Induction motors are an excellent choice for industrial motors due to their rugged construction, absence of brushes (which are required in most DC motors). With new modern power electronics, the speeds of the motors are controllable.

Concept Reinforcement

1. What is a stator?

2. What is a rotor?

3. List the three main types of three-phase motors.

Section 3.13 – U.S. and International Electrical Standards

Section Objective

- Explain US and international electrical standards

Purpose

The purpose of this section is to provide you with an understanding of the differences of electricity provided in different countries around the world. You will also learn what you need if you want to use your electrical devices in another country.

Electricity in Different Countries

If you have the opportunity to travel to a different country, depending on what country you are from and which country you would be visiting, you may find that your hair dryer won't work. The plug will not fit into the electrical outlet in the bathroom. The charger for your cell phone and computer may also be incompatible.

The world does not have one standard for all things electrical. As a matter of fact, there are several "standards" throughout the world.

What is different? The first difference you would see is the electrical outlet and the plug that goes into it.

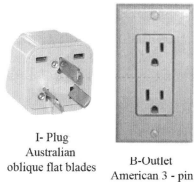

I- Plug
Australian
oblique flat blades

B-Outlet
American 3 - pin

Plug will not fit electrical outlet because they are from different countries.

Adapter

Electrical plug adapters are sold at most major international airports as well as some travel stores. Is the solution as simple as purchasing an adapter which converts the plug from one type to another?

The answer is: sometimes! It depends on the electrical device/appliance. While adapters may be purchased to change one type of plug to another, it is not as simple as converting the prongs in an electrical plug to fit another outlet. The power that is supplied to that outlet is not the same for each country. Before you plug in a device, you should make sure it is compatible. If the device is not compatible, it is possible that the device could be damaged.

The good news is that the units of measurement are universal. For example, current is measured in amperes or amps. Voltage is measured in volts. Frequency is measured in hertz (Hz). The concepts of direct current (DC) and alternating current (AC) are the same throughout the world. However, the voltage and frequency will vary from country to country.

Current (amperes or amps)

Current is measured in amperes or amps. The ampere (amp) is a measurement of the amount of electricity that flows through a circuit. An amp or ampere is represented by "I" or "A" in a formula. One ampere of current is defined when one coulomb (electrons) passes a point in a wire in one second. One coulomb equals 6.25×10^{18} electrons.

Current can be direct current (DC) or alternating current (AC). All countries supply alternating current for commercial and residential usage. A few places (such as Argentina, Bahrain and Johannesburg) provide both AC and DC in certain locations.

Voltage

The voltage that is delivered to an electrical outlet varies throughout the world. From 110 volts found in Japan to 240 volts found in Nigeria, the following voltages can also be found throughout the world: 110, 115, 120, 127, 220 and 230 volts. The United States uses primarily 120 volts and 220 volts.

Voltage is the force that pushes electrons through a wire. Voltage is the difference in electrical potential between two points in a circuit and is measured in volts (represented by "V" or "E"). A volt causes one coulomb to produce one joule of work.

Frequency

Alternating current (AC) changes the flow of electrons at a specific frequency. The unit of measurement is hertz (Hz). Hertz (Hz) is the number of alternating current (AC) cycles that occur each second. The good news is that there are only two "standards" in use. One is 50 Hz (50 cycles per second) and the other is 60 Hz (60 cycles per second). North America and parts of South America use 60 hertz (Hz). Most of the rest of the world uses 50 hertz (Hz). Even if voltages are the same, a 60-hertz clock will not function properly on a 50-hertz current nor will a 50-hertz clock function properly on a 60-hertz current.

Alternating current (AC) can be supplied in 1 phase, 2 phase or 3 phases. The conventional phases are single (1) phase and 3 phase.

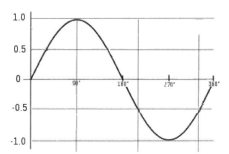

Single phase alternating current

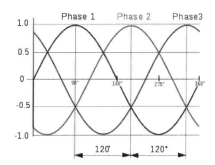

Three phase alternating current

Adaptors

The purpose of an adaptor is to change the physical nature of the plug so that it can be inserted into a different socket. The adaptor does not change the electrical voltage in any way. All it does is change the physical characteristics of the plug. Some small appliances are intended for international travel and have a switch which converts the voltage within the device. These devices have dual voltage capability. If the device cannot handle different voltages, you will probably need a voltage converter or transformer to make the device work.

Type of Plug	Description	Common User	Picture of Plug	Picture of Outlet
A	Flat blades	North American and Japanese		
B	Flat blades with round grounding pin	North American		
C	Round pin	European		
D	Round pins with ground	Older British		
E	Round pin (outlet contains grounding pin)	French		

238

F	"Schuko" (side grounding contacts)	Used in many countries.		
G	Rectangular blades	British		
H	Oblique flat blades with ground	Israeli		
I	Oblique flat blades	Australian		
J	Round pins with offset ground	Swiss		
K	Round pins with spade ground	Danish		
L	Round pins with ground	Italian 3-pin		
M	Round pins similar to type D but large plus one oversized pin	South African		

Converters and Transformers

Both converters and transformers step up or step down the voltage. Converters are lighter and less expensive than transformers and can be used short term on small electrical appliances that do not contain electronic chips.

Transformer to convert between 220/240 volts and 110/120 volts

A plug is no more than a pass through of current. Converters and transformers step down or step up voltage. The example shown above is a converter that converts European 220/240 volts to 110/120 volts. This enables a North American traveler to use small electronic items such as hair dryers, electric shavers, irons, coffee makers for short periods of time. This converter will support up to 1600 watts.

While transformers are sometimes more expensive and heavier than converters, they can be used to continuously operate electronic products containing such as radios, CD players, DVD players, computers and televisions.

Computers are electronic devices and should be used with a transformer, unless they are dual voltage. Today, most laptop battery chargers and AC adapters are manufactured with dual voltage and multiple frequencies. The back of the adapter will list any power that is supported. If the country provides one of the power choices listed on the back of the adapter, the computer will work with a compatible plug adapter for that country.

Computer charger showing support for different voltages and frequency

INPUT: AC 100V-240V~ 1.0A 50 / 60 Hz

Label on a computer charger showing support for different voltages and frequency

The label on this computer charger shows the range of voltage levels and different frequencies that are compatible. The computer charger is compatible with a "nominal" voltage anywhere from 100 volts to 240 volts and will operate when connected to a 50 hertz (Hz) or 60 hertz (Hz) power outlet. This computer charger can be plugged into the appropriate plug adaptor and used almost anywhere in the world. However, it is designed for alternating current only (AC). Some areas of the world do supply direct current (DC). Like most appliances, this computer charger should not be used on a direct current (DC) system.

If you measure the voltage at any given test point, it should be within a certain range of the expected value. The "nominal" or normal value is the expected or "average" value of the voltage. The actual nominal value is usually determined by calculating the square root of the average of the instantaneous voltages making up one cycle of the waveform. This is called the root mean square (RMS). Since voltage is an alternating current (AC) sine wave, the RMS value is used in order to give an equivalent of the power that would be developed if it was direct current (DC). Nominal voltage is the standard value used when referring to a voltage level. The actual voltage measured at any point in time could be 5 to 10 % higher or lower.

The transformer should be selected based on the actual amount of wattage needed to support the appliance. The manufacturer will specify the wattage required to operate an appliance. The transformer should always be larger than the manufacturer's specifications for the appliance. An additional 25% to 50% should always be added as a safety buffer to deal with variations in the power supplied and heat build-up in the transformer. To be safe on the safe side, use 50% in your calculations.

Example 1: The specifications for a microcassette transcriber are 120V, 60 Hz with a power consumption of 12 W using a type B plug (American 3 pin: 2 flat pins and 1 round grounding pin). What are the requirements to operate this device in Switzerland?

Note: Switzerland uses J plugs at 230V and 50 Hz alternating current (AC).

The physical connection must be made by converting the B plug to a J plug using an adaptor.

A transformer should be used to step up the power from 120V to 230V.

The transformer should be sized to provide 12 watts of power plus 50% for a total of 12 watts + 6 watts = 18 watts.

Since the transformer does not convert the frequency, the unit may run a little slower on 50 Hz, but it should still work. The quality of the sound may or may not be acceptable.

Example 2: If you use a power strip for multiple devices, you must calculate the combined wattage of all of the appliances that you plan to plug and add 50% to that total. Calculate the power requirements for a power strip with one computer at 90 watts and one radio at 300 watts.

The total power required would be:

(90 + 300) = 390 watts

This is the minimum power required. To properly size the transformer, you should add 50% to the total.

Calculate the safety factor of 50%:

0.5 × 390 watts = 195 watts

Add the safety factor to the minimum power requirements:

390 watts + 195 watts = 585 watts

The total amount of power this transformer must support is 585 watts. A 600 watt transformer would be acceptable.

In most cases, the manufacturer lists the voltage requirements (V) and power consumption (watts) for the device. If the wattage is not listed, but the voltage (V) and amperage (A) is provided you can calculate the wattage by multiplying the voltage by the amperage rating.

Volts × amps = watts

Both converters and transformers convert voltage (V) but not frequency (Hz). Some electronic devices are not adversely affected by the frequency difference as long as the voltage is corrected and the proper adaptor used to make the connection. Battery chargers usually fall into this category. They are not affected by the faster/slower cycles, but will take less/more time to charge.

The Difference between Converters and Transformers

Alternating current (AC) power is supplied in alternating bursts in the shape of a sine wave. To reduce 230V to 120V, the converter cuts the sine waves in half. The transformer actually alters the amplitude of the waves. This is the major difference between how converters alter voltage and transformers alter voltage.

Electronic devices require a full sine wave to function properly whereas electric appliances can function with either a full or a half sine wave. This means that electric appliances should work with either a converter or a transformer but electronic devices will only work with a transformer.

Summary

The world does not have a universal standard that is applied to electrical power. The one constant is the definitions used to define electrical power. While most countries supply alternating current, there are a few places where direct current is supplied. The nominal voltage and frequency vary between countries and are determined in each country by its own infrastructure (electrical utilities). The physical connectors between electrical devices and the supplied power source also vary between countries.

When traveling from one country into another, it is important to be aware of the differences if you plan to bring any electrical device with you. The infrastructure electrical specifications for each country are available and can be accessed via the Internet. Manufacturers supply the electrical specifications for electrical and electronic devices.

Concept Reinforcement

1. What is the definition of an electrical plug adaptor?

2. What is the major difference between a converter and a transformer?

3. What is the equation to calculate power if you are provided the voltage and the amperage of an electrical device?

4. True or False. A transformer steps up and/or steps down voltage when it is used to connect an electrical device into an electrical outlet.

Section 3.14 – How Things Work -Power Transmission and Alternative Sources

Section Objective

- Describe power transmission and alternative sources of power

Purpose

The purpose of this section is to gain a basic understanding of how power is distributed and why it is more economical to transmit alternating current power than direct current power over long distances. Various alternative power sources will also be discussed. The purpose of this section is to provide you with a basic understanding of the electrical environment in which you live.

A Little History

In **1831-1832** English chemist and physicist Michael Faraday discovered the operating principle of electromagnetic generators. **Faraday's Law states the electromagnetic force (EMF) induced in a circuit is directly proportional to the time rate of change of magnetic flux through the circuit**. In other words, any change in the magnetic environment of a coil of wire will cause a voltage (EMF) to be "induced" in the coil. Faraday also built the first electromagnetic generator, called the 'Faraday disc'. The generator used a copper disc rotating inside a horseshoe magnet. It produced direct current (DC).

In **1832** French instrument maker Hippolyte Pixii built the first practical mechanical generator of electrical current. He based his work on the principle of magnetic induction discovered by Michael Faraday.

At 3 p.m. on **September 4, 1882**, Thomas Edison's electric illuminating station using direct current (DC) went into operation on Pearl Street in New York City. One month later (**October 1982**), The United States Electric Illuminating Company of Charleston started up South Carolina's first central station for incandescent lighting. Less than one year later in July **1883** a three-wire system went into operation in Sunbury, Pennsylvania. **November 30, 1886** (only four years after Edison's Pearl Street Station opened) the first alternating current central station to operate commercially in the United States was placed in service in Buffalo, New York.

Nikola Tesla arrived in this country in 1884 with a letter of introduction to Thomas Edison, who hired him. Because Tesla was a strong believer in alternating current, he was in conflict with Edison. He left Edison's employment over a disagreement on pay. Later he managed to gain support from George Westinghouse. It was at the **1893** World's Fair that Tesla and George Westinghouse introduced visitors to the alternating current (AC) power they used to illuminate the Exposition. The "War of the Currents" between Edison and Tesla and Westinghouse was a conflict between Edison's direct current (DC) to provide power to everyone and Tesla's alternating current (AC). Tesla due to his eccentric personality and a number of scientific claims, was ultimately ostracized. Later recognition came for his extensive and revolutionary contributions. The tesla, the unit for measuring magnetic induction (magnetic field), was named in his honor.

A Quick Review

- **Current** is the flow of electrons along a path. An **electrical circuit** is a complete path along which electrons flow with a return to the source. In **alternating current**, the direction of the flow of electrons is reversed at regular intervals.

- **Electromagnetic force (EMF)** is the electrical pressure that pushed electrons through a conductor.

- By convention, the **direction of the magnetic field** is from N to S.

- By convention, the **direction of the flow of electric current** goes from plus (+) to minus (-).

- **Negatively charged electrons move from (-) to (+).** Electrons move in the opposite direction than the convention for the direction of current.

- The **ampere** (amp) is a measurement of the amount of electricity that flows through a circuit. An ampere or amp is represented by "I" or "A".

- **Voltage** is the electrical pressure which pushes electrons through an electric circuit. Voltage is the difference in electrical potential between two points in a circuit and is measured in volts. Volts are represented by "V" or "E."

- **Voltage drop** is the difference in voltage from one end of an electrical circuit to the other.

- **Wattage** is the amount of power that is used in an electric circuit. The unit of measurement is the watt and is represented by "W" or "P." One watt is one ampere of current flowing at one volt.

- **Power** is the rate at which work is performed. Electric power is the rate electrical energy is transferred by an electric circuit. Units of measurement are watts "W" and horsepower "hp". (1 horsepower = 746 watts) (1 watt = 0.0134 horsepower)

- **Work** is power used over a period of time. Units of measurement are joules, BTUs and calories. (1 BTU = 1050 joules) (1 calorie = 4.186 joules)

- Work and energy are basically the same and use the same units of measurement (joules, BTUs, and calories).

Definitions Useful to Explain Generation and Transmission of Power

- A **dynamo** is a stationary structure which provides a constant magnetic field by using electromagnets (field coils). The field coils are windings which rotate within the magnetic field.

- A **rolling blackout** (load shedding) is a scheduled electrical power outage. If the demand for power exceeds the electrical utility company's ability to produce electricity, the electrical company would rotate the power to its users for specific periods of time. In some countries rolling blackouts are standard. The schedules are fixed and people work around the known interruption times.

- **Electric power transmission** is the transfer of electrical power from the power plant to the consumer.

- The power distribution **grid** is the network of power transmissions that distribute power from its source to the consumer.

- Electrical **infrastructure** can be defined as the basic physical and organizational structures needed for the operation of the services and facilities necessary to provide electricity to the consumer.

- A **power outage** (power failure, power loss, or blackout) means short-term or long-term loss of power to an area that previously had electrical power. Power outages are categorized into three different phenomena, relating to the duration and effect of the outage. They are dropouts, brownouts, and blackouts.

- A **dropout** is very short loss of power (milliseconds to seconds) typically caused by a temporary fault on the power line. Power is usually restored once the fault is cleared. Flashing numbers on your digital clock is an indication that a loss of power occurred.

- If your lights dim, this is a drop in voltage and is called a **brownout.** This is not good for electric motors, so if you have any running it is best to shut them off until full power is restored.

- A **blackout** is a total loss of power to an area. Check your neighbor's homes. If they have power, check your circuit breakers. If your neighbors are without power, contact your local electric utility to report the outage. It may be something local that can be repaired quickly. If it is caused by a major storm, it may be awhile before power is restored. In that case, get out the flashlights and prepare to relax or take a nap.

- A **transformer** is a device that transfers electrical energy from one circuit to another. The purpose of a transformer is to step-up or step-down the voltage. Transformers are necessary for transmission of power over long distances.

- **Power quality** is a set of acceptable tolerances for electrical systems. Power supplied needs to function within the tolerances for the safety of people and equipment. In reality it is the quality of the voltage that is important.

- **Standby generators** automatically start up and supply electrical power when a power outage occurs. These are critical to support sites that are necessary for public safety (hospitals, telecommunications for police and fire departments, etc.)

- **UPS systems** (uninterruptible power supply) provide a constant flow of electricity in the event that a primary power supply becomes unavailable for a short period of time. These systems are usually installed to allow the orderly shutdown of computer systems and other electronic storage devices that are susceptible to data loss in the event of sudden loss of power.

Building Blocks for Transmission of Electricity

These are the requirements to create and transmit electricity. For the purpose of this lesson, a generic list is provided.

Energy source – the raw material used to provide the means to generate electricity. Energy sources include fossil fuels (coal, gas, and oil), nuclear power, hydroelectric power, and biomass. Wind power turns the rotor on the generator directly. The flow of water may also turn the rotor on the generator.

Boiler – converts the raw material to steam.

Turbine – steam powers the turbine which turns the rotor in a generator.

Generator – powered by the turbine, converts mechanical energy into electrical energy using electromagnetic induction. A coil of wires made of conducting material is rotated through a magnetic field. When a wire made of conducting material cuts through a magnetic field, an electric current is created in the wire.

Electrical substation–a subsidiary station in an electrical distribution system. The substation houses the transformers that step- up the voltage.

Step-up transformer – steps up the voltage while decreasing the current. This is in preparation for long distance transmission.

Transmission lines – transmits up to 1 million volts over long distance lines.

Transmission substation–connects two or more transmission lines. If the voltages are the same, high voltage switches are installed to allow connection as well as isolation for maintenance. Some transmission stations contain transformers to convert between two transmission voltages.

Step-down transformer – steps down the voltage while increasing the current. This is in preparation for use by consumers.

Electrical substation–a subsidiary station in an electrical distribution system. The substation houses the transformers that step- down the voltage.

Distribution substation – steps-down power for local distribution.

Local power lines – carry power from distribution substation to distribution transformer.

Distribution transformer – mounted near the end user for final step-down of voltage for end user.

Power line to electric meter – carry power from the distribution transformer and to the electric meter for use by the end user.

Power panel – receives power and distributes power as wired in the building or home of the end user.

Electric outlet – receives power from the power panel and is available for use.

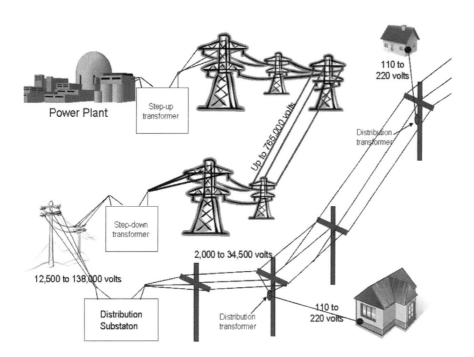

Power Generation, Transmission and Distribution

How Power is Transmitted

There are three parts to transmitting power. First is the generation of electricity. Next is the transmission to the desired destination. And third is the distribution to the consumer.

The energy source

Electricity is generated at a power station. The most common means to provide electricity is the burning of fossil fuels. The fossil fuel is burned in a boiler producing steam. The steam provides the power to drive a turbine. The turbine then turns the rotor in a generator.

The conversion to electricity

The electrical generator converts mechanical power into electrical energy. Three sets of alternating electric currents are produced from three coils (windings) of the generator. The currents produced are sinusoidal wave forms at the same frequency with the same amplitude. However the windings are designed and arranged so that the currents are offset by one-third of their cycle. In other words, they have a phase separation of 120°. The generator frequency is either 50 Hz or 60 Hz, depending on the country.

Note: there are a few places in the world that generate direct current (DC) power. This lesson will focus on the generation and transmission of three-phase alternating current (AC).

Step-up transformer in preparation for transmission

Once the current is generated, it flows through conductive wiring to a step-up transformer. The purpose of this transformer is to increase the voltage and decrease the current while maintaining the three-phase sinusoidal alternating current.

The step-up transformer consists of three magnetic steel cores with 2 copper wired windings on each core. Refer to the image. The windings are electrically insulated from one anther. The voltage source causes an alternating current (AC) to flow in the primary winding. The alternating current in the primary winding produces a magnetic field in the transformer core. The changing magnetic field expands and collapses cyclically. The magnetic field cuts through the wires of the secondary winding. The changing magnetic field induces a voltage in the secondary winding (Faraday's Law).

Faraday's Law states the electromagnetic force (EMF) induced in a circuit is directly proportional to the time rate of change of magnetic flux through the circuit.

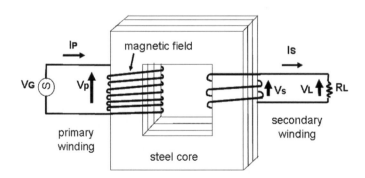

Simplified Transformer Model

Transmission over long distances

This step-up voltage is ready to be transmitted over transmission lines. Electricity is transmitted at very high voltages (ranges from 110,000 volts on up close to a million volts). The electricity being transmitted has high voltage and low amperage. Lower amperage loses less energy to the resistance of the wire over longer transmissions.

Step-down transformer in preparation for distribution

The purpose of this transformer is to decrease the voltage and increase the current in preparation for distribution.

The step-down transformer consists of three magnetic steel cores with 2 copper wired windings on each core. This is the reverse in operation of the step-up transformer.

Stepping down of the voltage is usually done in stages depending on distribution and the end user. Some manufacturing facilities require large amounts of electricity and they may provide their own substations.

Distribution substation

A distribution substation usually has at least two transmission lines. That allows for maintenance because part of the job of the distribution substation is to trouble shoot problems. The distribution substation also provides a step-down transformer. Distribution voltages are lower than the distribution station receives. The output is a number of feeders to distribution points.

Distribution

The feeders from the distribution substation are wires that run overhead along streets to distribution transformers located at or near the consumer. The final voltage step-down is provided by these local transformers.

The wires from the local transformers are wired directly to the electric meters, which monitor the consumption of power by the end user.

End User

Once power is supplied to an end user, it is the responsibility of the home owner or business owner to maintain the electrical wiring inside the building. It is up to the builder to make sure that the electrical safety standards are met.

Alternate Sources of Power

The majority of power is generated as a result of the conversion of fossil fuel and nuclear reactors. Solar energy, hydroelectric (water), natural gas, petroleum, wind, tidal harnesses, and geothermal all provide sources for the generation of electricity.

With the exception of photovoltaic panels, all of these sources do one thing. They provide the means to drive a turbine that drives a generator that produces electricity.

In a **coal** plant, steam is generated by burning coal in a boiler. The steam spins a turbine which has a generator connected to the same shaft.

Burning **oil (petroleum)** can also generate electricity. Oil fired turbines or oil fired boilers convert their energy by producing steam. The steam spins a turbine which has a generator connected to the same shaft.

Gas turbines are driven directly by gases produced by the combustion of **natural gas**. Combined gas turbine plants are driven by both steam and natural gas. They generate power by burning natural gas in a gas turbine and use residual heat to generate additional electricity from steam.

Enriched uranium (**nuclear**) is placed in a pool of cooling water. As the radioactivity is allowed to increase, the water will boil and can be used to generate steam through a heat exchanger. The steam spins a turbine which in turn drives a generator.

Solar furnaces collect sunlight and concentrate it on a container of water, turning it to steam. The steam turns a turbine generating electricity.

Photovoltaic panels convert sunlight directly to electricity. Photovoltaic cells are semi-conductors in which current flows when they are illuminated.

Wind mills are simple devices. The vanes catch the wind and turn. When they turn, they rotate the rotor of a generator.

The **hydroelectric** plant takes advantage of the fact that water flows downhill. A dam is constructed across a river and the water is allowed to pass through a turbine to get from the high side of the dam to the low side of the dam. This concept is also used to harness energy from the ocean. Underwater turbines are powered by the flow of the current. The blades of the turbines are turned by the flow of water (**tidal forces**); very much like the wind turbines are turned by the wind.

Geothermal power uses heat that is stored in the earth. Geothermal energy comes from radioactive decay of minerals and from solar energy absorbed at the surface.

Biomass uses living and recently dead biological material as fuel. The list of materials includes trash, dead trees, yard clippings, wood chips, biodegradable wastes and anything else that can be burnt as fuel. It does not include fossil fuels.

Alternative sources for electrical power

Summary

Electrical energy is produced by converting an energy source into mechanical power. One way to accomplish this is to burn fossil fuel. A generator converts the mechanical power into electricity.

In order to transmit electrical current over long distances, a transformer is used. The transformer transfers electrical power from one voltage to another. When a transformer steps up the voltage, it decreases the amperage (Faraday's Law).

It is more economical to transfer lower amperage electricity over long transmission lines because the diameter of the wire can be smaller. The smaller wire provides less resistance to the flow of current. Over long distances, the heat loss would be considerable in larger wires.

At the end of the transmission line (the receiving end), a step-down transformer is used to lower the voltage and increase the amperage. Because the voltage used for long distance transmission is so high, this step-down process can be done in stages.

As consumers increase the demand for more and more electricity, alternate sources are being developed. The ones most common today are fossil fuel and nuclear. Hydroelectric (water) power has also been around for a long time.

While wind power has powered windmills for ages, wind power is now being used to turn the rotor in generators and create electricity. Other sources of energy include solar, photovoltaic, tidal forces (flowing water), geothermal and biomass.

Concept Reinforcement

1. What is the primary purpose of a step-up transformer?

2. The electrical generator converts _____ power into electrical energy.

3. What is a brownout?

Section 3.15 – Applications of Electrical Engineering

Section Objective

- Describe practical applications of electrical engineering

Purpose

The purpose of this section is to review the concepts and applications of electrical engineering.

A Quick Review

- **Alternating current (AC)** is current that switches direction periodically. The movement (or flow) of electrons periodically reverses direction. Most power supplied to homes and businesses is alternating current.

- The **ampere** (amp) is a measurement of the amount of electricity that flows through a circuit. An ampere or amp is represented by "I" or "A".

- A **battery** is an energy source. It has two terminals, one positive and one negative. When the terminals are connected to an electric circuit, the voltage is the force that pushes the electrons through the conductor. The amount of voltage is measured in volts. When the batteries are connected in series, the total voltage is the sum of the individual voltages.

- **BTU** (British Thermal Unit) is a unit that measures heat (energy).

- **Direct current (DC)** is current that flows only in one direction from the energy source and back. Batteries provide direct current.

- By convention, the **direction of the magnetic field** is from N to S.

- By convention, the **direction of the flow of electric current** goes from plus (+) to minus (-).

- An **electric circuit** contains a power source, a conductor to carry electric current and a resistance (load). An **electric circuit** in a portable light consists of an energy source, a conductor to carry the electric current (flow of electrons) and a light source. Portable light electric circuits are direct current (current flows in one direction only).

- **Electrical energy** is measured in watt-hours. A more common term for energy measurement is kilowatt-hour (KWH). 1 kilowatt = 1000 watt-hours

- **Electrical power** is measured in watts. A more common term of power measurement is Megawatts (MW). 1 Megawatt = 1 million watts

- **Electric current (I)** is the flow of electrons along a conductor.

- **Electromagnetic force (EMF)** is the electrical pressure that pushes electrons through a conductor.

- An **energy source (V)** provides force to move the electrons along a conductor.

- An **electric switch** provides a means to close or open an electric circuit (starts or stops an electrical device).

- **Heat** is generated when electrons come upon a resistance that impedes their flow.

- **Heating capacity** is measured in **British Thermal Units (BTUs).**

- **An ion** is an atom that has lost or gained an electron and therefore has a positive or negative charge. Ions are drawn to oppositely charged areas. If the ion is negative (it has gained an electron), it moves towards a positive area. If the ion is positive (it has lost an electron), it moves towards a negative area. **Electric current** is also carried by the flow of ions.

- **Light** is composed of particles called photons. A **photon** is an elementary particle and is pure energy containing no mass

- A **light source** consumes energy and gives off light and heat. The amount of light and heat depends on the type of handheld device. The purpose of a portable light is to provide light. Any heat given off is considered energy wasted.

- **Negatively charged electrons move from (-) to (+).** Electrons move in the opposite direction than the convention for the direction of current.

- **Ohm's Law** states that the amount of steady current through a material is directly proportional to the voltage across the material, for some fixed temperature. **V = I × R**

- **Power** is the rate at which work is performed. Electric power is the rate electrical energy is transferred by an electric circuit. Units of measurement are watts "W" and horsepower "hp". (1 horsepower = 746 watts) (1 watt = 0.0134 horsepower)

- A **resistance (R)** consumes energy and is the load on a system.

- A **schematic** is an electrical blueprint (drawing) of a circuit design

- A **series circuit** is where the current flows only in one path. A **parallel circuit** is where the current flows in more than one path.

- **Valence electrons** are electrons that travel in the outer orbit of an atom. When the atom is bombarded, it gives up its valence electrons. The maximum number of valence electrons in an outer orbit is 8. The fewer the valence electrons in the outer orbit, the more likely the atom is to give them up. **Electric current** in carried by the flow of electrons through a conductor from a negatively charged area to a positively charged area. The amount of current is the number of these electrons flowing in a path in a conductor and is measured in amperes (amps).

- **Voltage** is the electrical pressure which pushes electrons through an electric circuit. Voltage is the difference in electrical potential between two points in a circuit and is measured in volts. Volts are represented by "V" or "E."

- **Voltage drop** is the difference in voltage from one end of an electrical circuit to the other

- **Wattage ratings** can be converted to BTUs consumed per hour by multiplying the number of watts by 3.413 (the number of BTUs/hour equaling one watt).

- **Wattage** is the amount of power that is used in an electric circuit. The unit of measurement is the watt and is represented by "W" or "P." One watt is one ampere of current flowing at one volt. **Watts** measure power (energy per unit time). 1 watt = 3.412 BTU/hour.

- **Work** is power used over a period of time. Units of measurement are joules, BTUs and calories. (1 BTU = 1050 joules) (1 calorie = 4.186 joules)

Note: Work and energy are basically the same and use the same units of measurement (joules, BTUs, and calories).

Applications of Electrical Engineering

On/Off Switches

The primary purpose of a switch is to provide a means to shut off power to electrical devices. By opening a circuit, a switch stops the flow of electrons. Any electrical device in that circuit is "shut off". No flow of electrons means there is no current and that means there is no power to operate electric devices.

Light Bulbs and Lighting

In new buildings, the time spent analyzing and designing the appropriate lighting for both function and aesthetics is well worth the effort. Lighting installations should look good and work well while taking advantage of the newer less energy consuming lights. There is an entire industry built around lighting. Remember to look at LED lighting as an alternative.

Incandescent and fluorescent light bulbs

Flashlights and Portable Lanterns

All portable lights are in an enclosed container with a light, an on/off switch, conductors to complete the circuit and an energy source. The energy source is usually a battery, but there are now hand held lights that can be charged through mechanical motion (your labor builds the energy). Multiple batteries are in series to provide more voltage to the light. Portable lights are simple in design but valuable in many ways.

Portable battery operated lantern

Car Batteries

The lead-acid battery in internal combustion engines has a short life time and disposal issues. Under certain circumstances, it can be dangerous to handle. Lead-acid batteries have been known to explode. And certainly the spillage of sulphuric acid can be very damaging.

A gasoline engine has fuel lines, exhaust pipes, coolant hoses and intake manifolds and looks like a plumbing project. An electric car is a wiring project. A hybrid is both. As gasoline prices fluctuate and environmental concerns increase, the pressure is on to find alternatives to the internal combustion engine.

Car battery

Water Pump in an Internal Combustion Engine of an Automobile

The water pump pumps coolant through the engine and keeps the engine from overheating. The cooling is necessary because when the fuel/air mixture burns and expands, it reaches very high temperatures. The water pump is indirectly powered by the electrical system in the car. It is actually operated by the belt attached to the crankshaft of the engine.

The crankshaft is initially turned by energy from the battery when the ignition system is activated by the key. The crankshaft starts the up and down motion of the pistons. Once the pistons are set in motion and the fuel/air mixture is ignited by sparks from the spark plugs, the pistons will continue their up and down motion based on energy derived from the burning/expansion of the fuel/air mixture.

The ignition switch remains as a closed switch in the electric circuit which is tied into the battery and alternator.

Once the pistons operate on the fuel/air mixture, their continued up and down motion takes over the turning of the crankshaft. The pistons will continue to operate (and the engine will run) as long as there is fuel pumped from the fuel tank, the ignition switch remains in the on position providing a closed circuit (the spark plugs need that spark to ignite the fuel), the cooling system cools the engine and nothing major goes wrong with the car.

Automobile water pump

Ceiling Fans

Ceiling fans can effectively reduce both cooling (summer) and heating (winter) costs by improving the circulation of air. In the summer, the ceiling fan blows air downward providing a breeze that acts like a wind chill increasing the comfort level even though the actual temperature is not affected by the ceiling fan.

In the winter the direction of the rotation of the ceiling fan is reversed. The fan circulates the warm air that rises to the ceiling back down into the living area providing even heating and comfort.

Fan blows air down.	Fan draws cool air up.
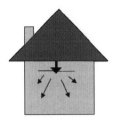	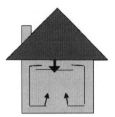
In summer, the downward movement of air creates a cooling "wind chill" effect. A room can actually feel several degrees cooler, without setting the thermostat lower.	In winter, hot air rises to the ceiling while cool air settles to the floor. Trapped against the ceiling, the warm air is wasted. Circulating the fan in reverse, moves warm air down from the ceiling into the living area and draws cool air up.

Ceiling fans can reduce energy costs.

Electric Heaters

Fueled heaters work well because they use little to no electricity, but these types of heaters cannot be used in areas without adequate ventilation. Portable electric space heaters work well in small closed areas. Electricity is available almost everywhere and electric heaters do not require any fuel to be stored.

Electric heaters are safe as long as safe operating procedures are used. Some are more efficient than others and there are many choices. Some are decorative and some are not.

Convection electric heaters use electricity to heat a resistor (element). The heat built up in the element is transferred into the area being heated. The method of transfer depends on the type of heater. Convection heaters use fans to blow hot air across the element. The air is warmed by the element and then blown into the room. Some units do not have a fan.

Infrared heaters use electromagnetic waves to bombard objects in the room thus heating them.

Heaters come with a variety of options, but some of the safety ones are most important. Automatic shutoff if overheating is very important. Automatic shutoff if tipping of the unit occurs is extremely important.

Heaters should not be placed near anything inflammable. Electric cords and plugs should be periodically inspected for wear.

Electric heaters if used properly can provide temporary energy efficient solutions to cold.

Portable electric heater.

Microwaves

Microwaves are very high frequency radio waves that swing back and forth (change electromagnetic fields from positive to negative) at a frequency of 2 billion cycles per second. Microwave ovens use microwaves to produce exactly the right wavelength to excite water molecules. Excited water molecules move faster and increase in temperature. Most food contains a significant amount of water. When the water molecules heat up and rotate, they bump into other molecules inside the food. The process is like frictional heating. Microwave energy converts to heat energy by causing food molecules to move faster. This heats the food product. The larger the proportion of water in the food, the faster the food product will heat up.

Microwaves usually cost less to cook food than a conventional oven. However, the biggest advantage is the speed at which the food is defrosted and cooked.

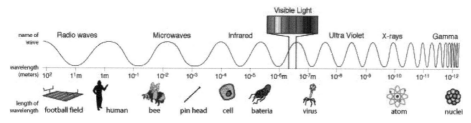

Microwave signals are between radio waves and infrared waves.

Wireless Remote Controllers

Two types of wireless remote controllers are very common. In each type of device, a hand held remote controller has buttons which correspond to circuits within the device. When a button is pushed, the associated circuit closes and an electromagnet signal is transmitted. The signal contains a series of binary pulses which are in fact a coded message. The coded message contains an identifier for the device that is to receive the message. It also contains a command message for the device. The device has a receiver that is tied into electrical circuits. When the receiver receives the message, it decodes the message. If it passes the identification test, then it decodes the command message. The electrical circuit in the device is preprogrammed to take a specific action for each command that it receives. Once it decodes a legitimate command, it will perform that specific action.

The difference between the two types of wireless remote controllers is the frequency of the electromagnetic waves transmitted. A wireless remote radio controller (RC) transmits radio waves that correspond to the binary code for the button being pushed.

A wireless remote infrared (IR) controller uses the frequency of light to carry binary code for the button being pushed. The light signals are transmitted by turning the light on and off in code between the remote controller and the device that it is controlling. Infrared light is invisible to human eyes and is in the invisible portion of the electromagnetic spectrum.

In the case of the RC controller, the RC receiver decodes the radio waves and converts them into commands. In the case of the IR controller, the IR receiver decodes the light signals and converts them into commands.

The greatest advantage of radio signals is their distance range and the fact that the signals can go through most walls. That is why a lot of IF transmitters now use an RF to IR converter to extend the range of the infrared remote controller.

A universal remote could save a lot of space and confusion.

Safety Sensors (Carbon Monoxide Sensors)

Safety sensors are devices that receive a signal and provide a preprogrammed response. They can be complicated or simple depending upon the application. Safety sensors are designed and implemented in order to reduce illness, injuries and death. Bypassing safety sensors is never a good idea. If a safety sensor triggers an alarm, the most important thing is to identify the alarm condition and take the appropriate action. If it is a carbon monoxide warning, immediate evacuation is critical. If it is a smoke alarm, again immediate evacuation is critical. Then call for help.

It is probably a good idea to implement the simplest solutions to monitor unsafe conditions. Sound audible alarms, light lights, ring bells or whatever it takes to get someone's attention. Then rely on human action to clear out an area. Only expert help with appropriate equipment should go back and fix the problem.

Carbon monoxide monitors can save lives.

Direct Current Motors

A direct current motor converts electrical energy into mechanical energy. The direct current motor contains a rotating part called an armature. The armature is made of a conductive material. When current flows through the armature, it produces a magnetic field around it. The armature is physically between field windings which are also made of conductive material. When a current carrying wire passes between two conductors, it creates a magnetic field between the conductors. When current is applied to the armature, the armature produces a magnetic field around itself and at the same time creates a magnetic field between the field windings. One side of the armature is attracted to the field windings and the other side is repulsed. This causes the armature to rotate through the magnetic field.

The armature is attached to a shaft which turns when the armature turns (current flows through the armature). The shaft is attached to a load. **Turning (driving) the load is the purpose of the motor.** The strength of the torque (turning of the shaft by the motor), determines how much of a load the motor can drive.

Two things determine the amount of torque produced by a motor:

- The strength of the magnetic field produced by the pole pieces.
- The strength of the magnetic field of the armature.

There are three main types of direct current motors:

- **Shunt direct current motors**—the field wirings are high-resistance and are connected in parallel with the armature. Speed is easily controlled in these motors. The speed can be changed by varying the field resistance or voltage supplied.

- **Series direct current motors**—the field windings are in series with the armature. The speed of series motors is difficult to control. The series direct current motor will develop the highest starting torque. They are ideal for electrical devices that require a high starting torque such as an automobile starter.

- **Compound direct current motors**—combine the best of both shunt and series motors. The compound direct current motor will develop less torque than the series direct current motor but more than the shunt direct current motor. There are three basic types of compound direct current motors:

- Cumulative compound direct current motor–connected so its coils aid the magnetic fields of both the series field and armature.

- Differential compound direct current motor–connected so its polarity is reversed to the polarity of the armature.

- Compound interpole direct current motor–has interpoles connected in series with the armature and series windings to make the magnetic field stronger.

The efficiency of a motor can be calculated if you know the voltage supplied, the current drawn and the actual horsepower output of the motor. Using Ohm's Law to calculate the power in where power in = voltage × current, then efficiency can be calculated by:

Efficiency = (power out/power in) × 100

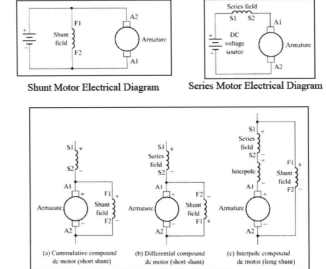

Direct current motors.

Alternating Current Motors

There are single-phase, split-phase and three-phase alternating motors. There are induction motors and non induction motors. Focusing on three-phase motors, there are three basic types:

- Squirrel-cage induction motor
- Wound Rotor Induction Motor
- Synchronous Motor

In each motor, there is a stationary part called a stator inside which are mounted poles. These poles are field windings of conductive material. They are spaced around the inside of the stator. Inside the stator is a rotor. A rotor is the part of the alternating current motor that rotates inside the stator. The rotor is also made of conductive material.

The one thing that makes all of the three-phase alternating current motors work is the rotating magnetic field.

The current is alternating current and it is also three-phase. The voltages supplied in a three-phase system are 120° out of phase with each other. The three voltages change polarity at regular intervals. The primary force of the magnetic field rotates as the power cycles.

The focus in this lesson was on the squirrel-cage induction motor. The squirrel-cage induction motor uses a rotor that consists of parallel bars with rings on each end.

Induction motors are an excellent choice for industrial motors due to their rugged construction, absence of brushes (which are required in most DC motors). With new modern power electronics, the speeds of the motors are controllable

Squirrel cage rotor

U.S. and International Electrical Standards

The world does not have a universal standard that is applied to electrical power. The one constant is the definitions used to define electrical power. While most countries supply alternating current, there are a few places where direct current is supplied. The nominal voltage and frequency vary between countries and are determined in each country by its own infrastructure (electrical utilities). The physical connectors between electrical devices and the supplied power source also vary between countries.

When traveling from one country into another, it is important to be aware of the differences if you plan to bring any electrical device with you. The infrastructure electrical specifications for each country are available and can be accessed via the Internet. Manufacturers supply the electrical specifications for electrical and electronic devices.

When the plug on the right doesn't fit the outlet in the middle, the converter on the left enables the connection.

Power Transmission and Alternative Energy Sources

Electrical energy is produced by converting an energy source into mechanical power. One way to accomplish this is to burn fossil fuel. A generator converts the mechanical power into electricity.

In order to transmit electrical current over long distances, a transformer is used. The transformer transfers electrical power from one voltage to another. When a transformer steps up the voltage, it decreases the amperage (Faraday's Law).

It is more economical to transfer lower amperage electricity over long transmission lines because the diameter of the wire can be smaller. The smaller wire provides less resistance to the flow of current. Over long distances the heat loss would be considerable in larger wires.

At the end of the transmission line (the receiving end), a step-down transformer is used to lower the voltage and increase the amperage. Because the voltage used for long distance transmission is so high, this step-down process can be done in stages.

As consumers increase the demand for more and more electricity, alternate sources are being developed. The ones most common today are fossil fuel and nuclear. Hydroelectric (water) power has also been around for a long time.

While wind power has powered wind mills for ages, wind power is now being used to turn the rotor in generators and create electricity. Other sources of energy include solar, photovoltaic, tidal forces (flowing water), geothermal and biomass.

Old fashioned windmill

Summary

Engineering is the practical application of technology. Electrical Engineering is the practical application of electricity. Electricity is so much a part of our daily lives that it is very difficult to do anything that does not involve the use or application of electrical power.

It is important to understand where electricity comes from and how it is used. It is also important to understand how electricity works. Each of the units covered a different application of electricity.

Don't forget Ohm's Law which states that the amount of steady current through a material is directly proportional to the voltage across the material, for some fixed temperature.

V = I × R

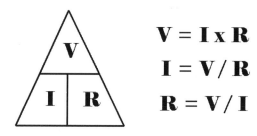

Ohm's Law and derivations

Concept Reinforcement

1. What is the difference between alternating current (AC) and direct current (DC)?

2. True or false. The electrical units of measurement (coulombs, watts, amperes, etc.) are standard throughout the world.

3. True or false. Direct current (DC) is the current transferred over long transmission lines.

4. Name four sources of energy used to generate electricity.

Appendix

Electrical Engineering Answer Key – Unit 1

Section 1.1

1. Student should start with waking up without an electric alarm clock; getting dressed without using anything that requires power such as a hair dryer; eating a cold breakfast; going to school where there are no electric lights, no working computer terminals, etc. and coming home to no TV or working internet.

2. If cold, would need some form of heater (wood or gas powered) and blankets. Lighting alternatives (flashlights + extra batteries, candles + matches, oil lamps + matches + extra oil). Food that doesn't require refrigeration or cooking (unless alternate means of cooking available such as gas grill). Battery operated radio for information. Bottled water.

3. Home: backup supplies including flashlights, batteries, non-perishable food items (canned goods – need can-opener), first aid kit, bottled water, blankets.

   School: flashlight, batteries, first aid kit, snack items, bottled water. Should always have food and water supplies, flashlights + batteries, candles + matches, portable lanterns + batteries, non-perishable food items and bottled water, first aid kit, battery operated radio, blankets, any critical medicines family member may require, charged cell phones, list of emergency phone numbers.

4. Walking, bicycling, horse riding, skating, skiing, animal drawn carriages or buggies, sailing.

Section 1.2

1. The Law of Charges states that opposite charges attract and like charges repel. The Law of Centrifugal Force states that a rotating or spinning object will pull away from its center the faster it spins. Centrifugal Force is an outward force associated with rotation. Either answer to Centrifugal Force is acceptable.

2. Protons, neutrons and electrons

3. Valence electrons are the electrons that reside in the outermost orbit (shell) around the nucleus.

4. When a valence electron is hit, it is knocked out of its orbit and moves on to hit another valence electron in an adjacent atom. This flow of negatively charged electrons is electric current.

Section 1.3

1. The ampere (amp) is a measurement of the amount of electricity that flows through a circuit. One ampere of current occurs when one coulomb of electrical charge passes a point in a wire in one second. Amp is an abbreviation of ampere. Any of the above answers is acceptable.

2. An ohm is the amount of resistance that allows 1 amp of current to flow when 1 volt is applied to the circuit. The resistance in the system is what consumes/reduces energy. One ohm is the resistance value through which one volt will maintain a current of one ampere. Any of the above answers is acceptable.

3. Voltage is the force that pushes electrons through a wire. Voltage is defined as electromotive force (EMF). A volt causes one coulomb to produce one joule of work. Any of the above answers is acceptable.

Section 1.4

1. Ohm's law states that the amount of steady current through a material is directly proportional to the voltage across the material, for some fixed temperature. Ohm's Law is: $V = I \times R$

   The two variations of Ohm's law are:
   $I = V/R$
   $R = V/I$

2. Conductors are used to carry electrons. Conductors carry electricity. Either answer is acceptable.

3. A generator converts mechanical energy into electrical energy.

4. 1. Solar or sun
   2. Hydroelectric or water
   3. Wind

   Other renewable energy sources would be biomass (burning of waste) and geothermal wells.

Section 1.5

1. Power is the rate at which work is performed. Power is the rate energy is used. Work is power used over a period of time. Any of the above answers is acceptable.

2. Joules, BTUs, calories, ft-lbs

3. Watts, horsepower

4. power = work / time
   or
   watts = BTU/hour
   horsepower = ft-lbs/second

Section 1.6

1. An electric circuit that contains a break in the path so no electrons will flow.

2. An electric circuit that provides a path for electrons to return to the source of power. There is no break in the circuit so electrons will flow.

3. A circuit breaker breaks the circuit. It creates an open circuit. It opens the circuit when there is an overload of current. It trips when there is an overload and opens the circuit. It is a safety device. All of the above answers are acceptable.

4. A fuse "breaks" when it overloads and must be replaced while a circuit breaker can be reset.

Section 1.7

1. A conductor provides a path for electric current. A conductor is made up of atoms with fewer electrons in their outer orbit and will easily give up their electrons. A conductor allows electrons to flow. A conductor is made up of atoms with fewer valence electrons which will move when an energy source is applied. Any of the above answers is acceptable.

2. An atom that holds onto its electrons is an insulator. Electrons have a difficult time moving through an insulator. An insulator protects a conductor. Any of the above answers is acceptable.

3. A resistor slows the flow of current. Resistors dissipate energy in the form of heat. A resistor slows the flow of electrons. A resistor is some form of load on an electric circuit. Any of the above answers is acceptable. A variable resistor has the ability to change the amount of resistance to the flow of electrons or current in an electrical circuit. |

4. A capacitor stores the potential energy of accumulated electrons in the form of an electric field. A capacitor is a form of storage device for electrons. Either answer is acceptable.

Section 1.8

1. A battery is a closed container full of chemicals that produce electrons. A battery is an independent power source providing direct electrical current. Either answer is acceptable.

2. When a battery is connected in a closed circuit, electrons will flow through the circuit. A closed circuit is an electric circuit that provides a path for electrons to return to the source of power. A closed circuit is an electric circuit with no break in the flow of electrons. Any of the above answers is acceptable.

3. Cell phone, laptop computer, hand-held video game, car, iPod, flashlight, travel alarm, portable radio, TV remote controller, portable CD player, portable DVD player, some garage door openers, digital camera. Student can add to this list.

Section 1.9

1. Direct current is a closed circuit with the current (electrons) flowing in one direction.

2. Alternating current is a closed circuit with bidirectional current flow. The direction of the flow of electrons is reversed at regular intervals. Either answer is acceptable.

3. A transformer is a device that can step up or step down the voltage.

Section 1.10

1. Static means stationary or non-movement.

2. When atoms lose or gain electrons, there is an imbalance: the atoms have a negative or positive charge. Static electricity is the result of movement of electrons from one object to another creating a negative or positive charge. Either answer is acceptable.

3. Discharge is when electrons move from an imbalanced object. Discharge is the reverse of the buildup of static electricity.

Section 1.11

1. The circuit is closed.

2. The circuit to the lamp is open.

3. The circuit to the lamp is open.

Section 1.12

1. A magnet produces a magnetic field around it.

2. Magnetite and lodestone.

3. When current is passed through a coil of conductive material, the coil behaves like a magnet.

Section 1.13

1. When electrons meet resistance, energy is dissipated into heat.

2. The amount of current is directly proportional to the voltage supplied and inversely proportional to the resistance. The flow of electrons is directly proportional to the power source and inversely proportional to the resistance. It takes one volt to push one amp of current through one ohm of resistance. $V = I \times R$. Any of the above answers is acceptable.

3. When electrons flow through the filament in a light bulb, the energy of electrons meet resistance. The energy is converted to heat and the heat in turn creates light.

Section 1.14

1. Grounding provides a safe outlet for the excess buildup of electrical charge. Ground is a direct link to the earth.

2. It is potentially dangerous.

3. The best choice would be to seek shelter indoors. If that is not possible, you would want to stay away from anything that would attract lightning such as a tree.

4. No

Section 1.15

1. An open circuit is an electrical path that has some form of interruption to the flow of electrons. The circuit is broken at some point between the source and the return to source.

2. A closed circuit is an electrical path along which electrons flow with no breaks or interruptions in the flow of electrons.

3. A circuit breaker is a safety device that will interrupt the flow of electrons.

4. When anyone is working on electrical wiring, the circuit breaker should be tripped manually to prevent the flow of electrons (current).

Electrical Engineering Answer Key – Unit 2

Section 2.1

1. The purpose of electronic symbols is to provide a means of communicating ideas or concepts in writing. The graphical representation of electronic symbols is used to convey ideas and to document designs. Electrical schematics use electronic symbols to provide the information needed to build anything electronic. Any of these answers is acceptable.

2. Electrical schematics are electrical drawings.

3. battery

   generator

   resistor

   capacitor

   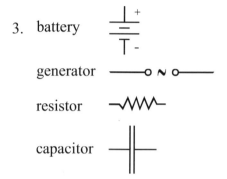

4. The flow of current is represented in an electrical schematic by a solid line. The solid line in a schematic represents a conductor which transmits current. Either answer is acceptable.

Section 2.2

1. Amperes or amps, either answer is acceptable.

2. Watt or horsepower, either answer is acceptable.

3. Ohms

4. $V = I \times R$
   Ohm's Law describes the relationships between voltage, resistance and current. It takes one volt to push one amp of current through one ohm of resistance. Ohm's Law states that the amount of steady current through a material is directly proportional to the voltage across the material for some fixed temperature. Any of the above answers is acceptable.

Section 2.3

1. Ohm's Law is: $V = I \times R$
   The two derivations of Ohm's Law are: $I = V/R$ and $R = V/I$

2. 1. There is only one path for current to flow.
   2. Current is the same at any point in the circuit.
   3. The total resistance is the sum of the individual resistors.
   4. The amount of voltage drops across all the resistors must equal the voltage applied to the circuit.

3. Current is the same at any point in the circuit:
   $I_{total} = I_1 = I_2 = I_3 = I_4 = 2$ amps
   $V_{total} = 96$ volts
   $R_1 = 20$ ohms
   $R_2 = 10$ ohms
   $R_3 = 16$ ohms
   $R_4 = 2$ ohms
   $V_1 = I_1 \times R_1 = 2$ amps $\times$ 20 ohms
   $V_1 = 40$ volts

   To calculate V_2, V_3 and V_4:
   $V_2 = I_2 \times R_2 = 2$ amps $\times$ 10 ohms
   $V_2 = 20$ volts
   $V_3 = I_3 \times R_3 = 2$ amps $\times$ 16 ohms
   $V_3 = 32$ volts
   $V_4 = I_4 \times R_4 = 2$ amps $\times$ 2 ohms
   $V_4 = 4$ volts

4. $V_{total} = V_1 + V_2 + V_3 + V_4$
   40 volts + 20 volts + 32 volts + 4 volts
   $V_{total} = 96$ volts
   The answer is yes.

Section 2.4

1. $V = I \times R$
   $I = V/R$
   $R = V/I$

2. 1. There are multiple paths for current to flow.
   2. The total current is the sum of the current that flows through all of the branches of the circuit.
   3. The voltage across any part of the circuit is the same as the applied voltage.
   4. The total resistance is the reciprocal of the sum of the reciprocals of the resistance of each branch of the circuit.

3. The voltage across any of the parallel branches is the same as the applied voltage so $V_1 = V_2 = V_3 = V_{total}$.

V_{total} can also be calculated if the total current and the individual resistors are known:
$V_{total} = I_{total} \times R_{total}$
where
$R_{total} = 1/(1/R_1 + 1/R_2 + 1/R_3)$
Either answer is acceptable.

Section 2.5

1. A combination circuit is a circuit that contains both series circuits and parallel circuits.

2. False. Ohm's Law always applies to any electric circuit.

3. 1. There is only one path for current to flow.
   2. Current is the same at any point in the circuit.
   3. The total resistance is the sum of the individual resistors.
   4. The amount of voltage drops across all the resistors must equal the voltage applied to the circuit.

4. 1. There are multiple paths for current to flow.
   2. The total current is the sum of the current that flows through all of the branches of the circuit.
   3. The voltage across any part of the circuit is the same as the applied voltage.
   4. The total resistance is the reciprocal of the sum of the reciprocals of the resistance of each branch of the circuit.

Section 2.6

1. A combination circuit is a circuit that contains both series circuits and parallel circuits.

2. False. Gustav Kirchhoff based his work on Georg Ohm's findings.

3. Kirchhoff's Current Law states that the sum of all current in a closed circuit is equal to zero. The current going into a node of a combination circuit is equal to the current leaving the node. Either answer is acceptable.

Section 2.7

1. A combination circuit contains both series circuits and parallel circuits.

2. Energy can never be created or destroyed within an isolated system.

3. Kirchhoff's Voltage Law states that the sum of the voltages around any closed circuit must equal zero. The voltage applied to a closed system is the total of the voltage drops on each of the resistors. The voltage drop across each resistor adds up to the voltage supplied by the source. The voltage applied to a closed circuit is equal to the sum of the voltage drops across all the resistors. Any of the above answers is acceptable.

Section 2.8

1. The three basic types of electronic circuits are series circuits, parallel circuits and combination circuits.

2. You would use a derivation of Ohm's Law to calculate the current. Ohm's Law is $V = I \times R$ therefore $I = V/R$ and in a series circuit R_{total} = sum of all the resistors.

3. When current reaches a node in a parallel segment of a circuit, the current splits between the parallel legs of the circuit. The amount of current that flows through each parallel segment depends upon the amount of resistance that is in each segment. When the current passes through the return node (where the parallel segments are joined to complete the return path), the current is combined to continue its return to source. The current is the same going into the parallel node as it is coming out of the return parallel node. It only splits during the time it flows through the individual parallel segments.

Section 2.9

1. The primary purpose of Thévenin's Theorem is to simplify complex networks. The theorem allows a complicated circuit containing many energy sources and many resistors to be represented by a single voltage source and a single resistor. Either answer is acceptable.

2. The black box represents a complicated circuit containing many energy sources and many resistors.

3. When Thévenin's Theorem is applied the resulting circuit is a "black box" that contains a single energy source and a single resistor.

Section 2.10

1. The primary purpose of Norton's Theorem is to simplify complex networks.

2. When Norton's Theorem is applied the resulting circuit is a "black box" that contains a single current source and a parallel resistor.

3. False. Thévenin's black box can easily be converted to a Norton equivalent circuit using Ohm's Law.

Section 2.11

1. Replace all but one of the voltage sources with a short. Replace all current sources with an "open". Solve for the current going through each resistor. Go back and follow the same procedure eliminating a different voltage source each time. When the current going through each of the resistors has been calculated for each voltage source, add up the current going through each resistor for each of the voltage sources.

2. True

3. The Superposition Theorem is a combination of both of Kirchhoff's Laws, Norton's Theorem and Thévenin's Theorem.

Section 2.12

1. False. Millman's Theorem provides a means to compute the voltage across parallel branches.

2. False. Millman's Theorem converts each branch in a circuit to its equivalent Thévenin's "black box" which he then converts to an equivalent Norton circuit. Once the segments are all parallel (Norton), Millman uses the parallel resistance formula to calculate the total voltage across all branches of the circuit.

3. $$V_{across\ all\ branches} = \frac{\frac{V_1}{R_1} + \frac{V_2}{R_2} + \frac{V_3}{R_3}}{\frac{1}{R_1} + \frac{1}{R_2} + \frac{1}{R_3}}$$

   V_1 = 24 volts
   V_2 = 12 volts
   V_3 = 12 volts
   R_1 = 10 ohms
   R_2 = 5 ohms
   R_3 = 20 ohms
   Substituting in Millman's equation,
   $V_{across\ all\ branches}$ = 15.428 volts

Section 2.13

1. The ohmmeter is used to measure ohms (resistance).

2. The ammeter is used to measure amps (current).

3. The voltmeter measures voltage (energy).

4. Analog meters use a pointer and scale to provide the measurement while digital meters use actual numbers on a display.

Section 2.14

1. False. A galvanometer is a type of ammeter and measures current.

2. True

3. The Wheatstone bridge contains three (3) known resistors.

4. The Wheatstone bridge contains one (1) unknown resistor.

Section 2.15

1. The three types of circuits are series circuits, parallel circuits and combination circuits.

2. resistor ─/\/\/\─

3. The ammeter measures the amount of current flowing through a conductor.

4. Superposition Theorem and Ohm's Law.

Electrical Engineering Answer Key – Unit 3

Section 3.1

1. The primary purpose of a switch in an electric circuit is to either turn on or shut off power to electrical devices. A switch opens or closes a circuit. Either answer is acceptable.

2. False. Switches will work in both alternating current circuits and direct current circuits.

3. When a switch is "on", the circuit should be closed allowing current to flow. However, be aware that sometimes switches are wired incorrectly so that the reverse is true.

4. An on/off switch stays in the state that is selected. An on/off momentary switch returns to its "normal" state when it is released.

Section 3.2

1. An incandescent light generates the most heat.

2. An ion is an atom that has lost or gained an electron providing the atom with a positive or negative charge. An electron is part of an atom and travels in an orbit or shell around the nucleus of the atom. The electron has a negative charge.

3. This type of light is usually a Light-emitting diode (LED).

4. The fluorescent light is more energy efficient than the incandescent light.

Section 3.3

1. Batteries in a standard flashlight are connected in series.

2. 1. A case that contains the entire device.
   2. An external on/off switch that turns the light on/off.
   3. A light source (incandescent bulb)
   4. A reflector that redirects the light rays from the light source, creating a steady beam of light.
   5. A lens to protect the light source (light bulb).
   6. A source of energy (batteries)
   7. A conductor that connects the energy source (batteries) to the light source and the on/off switch.

3. False. On/off switches are necessary in order to power down the portable light when not in use. If always left on, the batteries would soon be drained and no light would be available.

4. False. It does matter which way batteries are installed. If installed improperly, the portable light will not work.

Section 3.4

1. The standard lead acid car battery usually contains 6 battery cells.

2. Since the battery cells are connected in series, the total expected voltage is 6 times 2.1 volts for a total of 12.6 volts.

3. Turning the key in the ignition closes the circuit when starting a car.

4. The chemical interaction inside the battery creates voltage.

Section 3.5

1. The primary purpose of a pump is to move material from one place to another.

2. False. Some pumps are manual and can be operated without electricity.

3. The energy source required to start an internal combustion engine is a battery.

4. The ignition switch remains as a closed switch in the electric circuit which is tied into the battery and alternator.

Section 3.6

1. True

2. False. Running ceiling fans at the same time as running air conditioners can improve the efficiency of the air conditioner.

3. energy consumed = energy × time
   3000 watts × 24 hours/1000 = 72 kilowatt hours consumed in one day

4. energy consumed = energy × time
   90 watts × 24 hours/1000 = 2.16 kilowatt hours consumed in one day

Section 3.7

1. False. Infrared heaters use far-infrared waves to heat objects.

2. a. ceramic heater is a convection heater.

3. False. Electric heaters should be turned off when leaving the room for any length of time.

4. False. Estimating is not an exact science.

Section 3.8

1. The primary purpose of a microwave is to heat/reheat/defrost or cook food or beverages very quickly. Any of the above answers is acceptable.

2. False. Microwave us less energy than conventional ovens to cook food.

3. False. Metal containers should never be used in a microwave. Use of metal containers could damage the magnetron tube in the microwave.

4. energy consumed = energy × time
   1000 watts × 2/60 hours = 33.33 watt hours
   33.33 watt hours/1000 = 0.033 kWh

Section 3.9

1. The purpose of a wireless remote controller is to operate an electrical device from a distance.

2. radio remote controllers, infrared remote controllers

3. The radio remote controller works over longer distances. Radio signals can go through walls. Either answer is acceptable.

4. Remote controlled robotic devices, remote explosive triggering devices, remote operation of lights, and remote operation of overhead cranes. Any of the above answers or any similar answer is acceptable.

Section 3.10

1. A sensor is any device that receives a signal and responds with a specific action based on the signal it receives.

2. True

3. True

4. False. At 1600 PPM of continued exposure, death will probably occur within the hour.

Section 3.11

1. The three types of direct current motors are shunt direct current motor, series direct current motor, and compound direct current motor.

2. efficiency = (power out/power in) x 100

3. True

4. shunt

Section 3.12

1. A stator is the stationary part of an alternating current motor.

2. A rotor is the rotating part of an alternating current motor.

3. The three main types of three-phase motors are squirrel-cage induction motor, wound rotor induction motor, and synchronous motor.

Section 3.13

1. The purpose of an adaptor is to change the physical characteristics of a plug so that it will fit into a different socket (electric outlet). Different countries have different physical standards.

2. A converter cuts the sine wave of alternating current in half while the transformer keeps the full sine wave but alters the amplitude of the sine wave. Electric appliances will work with either a converter or transformer, but electronic devices will only work with a transformer. Either answer is acceptable.

3. Power = Current × Voltage
P = I × V

Section 3.14

1. The primary purpose of a step-up transformer is to increase voltage while decreasing the current. The step-up transformer prepares electricity to be transmitted over long distances. Either answer is acceptable.

2. mechanical

3. A drop in voltage is called a brownout. Lights dim but do not go out.

Section 3.15

1. The direction of flow of electrons in alternating current is periodically reversed. In direct current, electrons flow in one direction.

2. True

3. False. Alternating current is transferred over long transmissions lines.

4. coal, gas, oil, nuclear, hydroelectric (water), biomass, solar (sun), wind power, geothermal wells

Printed in Great Britain
by Amazon